In the Herbarium

In the Herbarium

The Hidden World of Collecting and Preserving Plants

Maura C. Flannery

Yale

UNIVERSITY PRESS

New Haven & London

Published with assistance from the Louis Stern Memorial Fund.

Yale University Press books may be purchased in quantity for educational, business,
or promotional use. For information, please e-mail sales.press@yale.edu (U.S. office)
or sales@yaleup.co.uk (U.K. office).

Set in Bulmer type by IDS Infotech, Ltd.
Printed in the United States of America.

Library of Congress Control Number: 2022942256
ISBN 978-0-300-24791-6 (hardcover : alk. paper)

A catalogue record for this book is available from the British Library.

This paper meets the requirements of ANSI/NISO Z39.48-1992
(Permanence of Paper).

10 9 8 7 6 5 4 3 2 1

To those I love who are not here to hold this book in their hands

Contents

Acknowledgments

MY PARENTS CATHERINE AND HENRY flannery taught me to love learning, and my husband Robert Hendrick taught me to love life and art. They definitely get most of the credit for this book. Then there are the untold joys given me by my sister Aideen Nolan and my stepson and his wife, Geoff and Laura Hendrick. Next come Elizabeth Michael Boyle, Lisa Getman, Cecily Cannan Selby, and Julia Upton, who have supported me in endless ways, as have many friends past and present.

St. John's University was my academic home throughout my career; its Faculty Writing Initiative was a significant source of support. Herrick Brown and John Nelson of the A. C. Moore Herbarium at the University of South Carolina have allowed me to learn from them, as have Nicole Tarnowsky and Kim Watson at New York Botanical Garden. I am enormously grateful to those in the herbaria, museums, colleges, universities, and libraries I've visited while researching this book. They were so generous with their time and expertise, both during visits and later when I needed more information.

The manuscript's reviewers worked very carefully and thoughtfully; their input helped me to greatly improve its content and style. Finally, Jean Thomson Black and Elizabeth Sylvia of Yale University Press were a wonderful blend of encouragement, patience, and advice.

Opening Hidden Gardens

THE ONLY WAY I CAN DESCRIBE HOW I came to write a book about herbaria is to say I was moonstruck. In 2010, while I was on a tour of the natural history museum in Providence, Rhode Island, the curator took us behind the scenes, showing us stuffed animals, fossils, and cabinets filled with folders containing large sheets of white paper mounted with pressed plants, the herbarium.[1] Then she pulled out albums of seaweed specimens from the nineteenth century. Two images remained with me when I left there: the information stored in those folders and the beauty enclosed in the albums. Though I am a biologist, before that tour I knew almost nothing about herbaria. After that I was in love with them, and like anyone in love I could think of little else.

Why do I love herbaria? First there is the history, dating back five hundred years, to even before the founding of the first botanical gardens. There are specimens collected in the Middle East in the sixteenth century, on Captain Cook's voyages around the world, and by such notables as Jean-Jacques Rousseau, Emily Dickinson, and Paul Klee. A single sheet can document a new species as well as its journey through time: for example, a plant collected by the American explorers Meriwether Lewis and William Clark (fig. I.1) was subsequently transported to Britain, where it was described by a German botanist, then owned by a British botanist, and ultimately sold at auction to an American, who donated it to the Academy of Natural Sciences in Philadelphia.[2]

It's fascinating to trace such stories, and there is a serious reason for doing so. Each of the millions of specimens stored in herbaria around the world records a species growing in a particular place at a particular time, each unique and irreplaceable. A preserved and labeled plant can reveal a

Figure I.1. Specimen of Oregon grape (*Berberis aquifolium*), with numerous determination slips verifying the name. It was collected on the Lewis and Clark Expedition at rapids along the Columbia River in Oregon. (Academy of Natural Sciences of Drexel University, ANSP Library & Archives PH Herbarium PH00043503)

great deal, not only about the species itself but also about its habitat, making it useful in studying ecology and environmental change. Also, DNA sequencing can be done on a small sample from it, providing information on its relationship to other species. Finally, in the twenty-first century, specimens are being digitized, making the information they contain available to a much broader audience geographically, including those in developing nations, where many of these plants were collected but then sent to herbaria in colonizers' home countries. This access is crucial to efforts to preserve biodiversity and deal with climate change. Herbarium sheets provide the ultimate evidence of a species' characteristics, and specimens from past centuries substantiate how flowering times and other traits have changed.

As I learned more about herbaria, I wanted to take a deeper dive into this world. I began to visit botanic gardens, museums, and universities, talking with those who care for collections as well as those who found and identified new species. I explored old herbaria hidden away in libraries and natural history museums, opened storage cabinets and smelled the aroma of mint plants collected decades ago, and even volunteered at herbaria. I learned to mount specimens and to photograph sheets and input label information into databases, so the image and related information could be uploaded to websites. I couldn't keep what I learned to myself so I began writing about herbaria, first in articles, then a blog, and now a book. My aim is to make the world of herbaria more widely known and to explore the many topics related to plant collecting.

The scientific worth of herbaria is at the fore, but there are many other ways of appraising them—as historical, aesthetic, cultural, and ethnobotanical collections. Deborah Harkness argues, "Every time a dried plant specimen changed hands it became infused with new cultural and intellectual currency as its provenance became richer, its associations greater."[3] Added notes, sketches, and name changes augment this richness. Value relates to the amount of information on the sheet, as present-day researchers well know when they search for data on flowering times and habitat.

Each collection is unique and a reflection of who created it: how the plants are pressed, what is collected, what data are recorded. Personal collections are particularly distinctive, not only in presentation style but also in what is chosen and what information is deemed important. The philosopher John Stuart Mill was interested in botany throughout his life, in part because he saw the hierarchical classification of living things as a model for ordering many aspects of human affairs, such as law. The Reverend William Keble Martin documented the plants he gathered in the English countryside with watercolors and an herbarium that became a reference in writing his book *The Concise British Flora in Colour,* published when he was eighty-eight years old. The twentieth-century composer John Cage was fascinated by fungi, while the Marxist revolutionary Rosa Luxemburg filled seventeen notebooks with specimens, mostly wildflowers, between 1913 and 1918, when she was jailed several times for her political activities in Germany. Her herbarium is how she kept in touch with what she considered the best parts of her world.[4]

For Carl Linnaeus, the great classifier, the herbarium was his major tool in organizing knowledge about plants. For amateurs, it is a learning device as well as a symbol of their seriousness in studying botany: their specimens are labeled because their owners are educated enough to identify species or were sufficiently connected in the botanical world to seek out identifications by others. When Jean-Jacques Rousseau was learning botany, his teacher had a "working herbarium," a rather rough and messy one that he referenced often in attempting to identify the plants he collected; it was not neat and well presented, but it served his purposes.[5] For botanical artists, the herbarium has provided references for plant structures and models for paintings. The plant itself, as it was for Linnaeus, was the final arbiter of accuracy.

For the colonial Philadelphia nurseryman John Bartram, his collection was a means to display his wares to his British patrons and also to help him learn botany when his London agent Peter Collinson identified his specimens. For British explorer and botanist Joseph Banks, collections could be instruments of economic and political power as colonizers sought new and profitable sources of food, medicine, fiber, and timber.[6] For many wealthy gardeners, an herbarium was a way to document what they grew— often exotics that spoke of wealth and status. For a World War I French soldier, collecting plants was a way to focus on something other than fighting, and for a Czech mother, a book of pressed plants was a remembrance of the botany excursions she had made with her son, something to tuck into his suitcase as he left in the Kindertransport before World War II.

There is a gestalt to herbarium specimens created by an individual, but it becomes obvious only after examining many sheets. For older specimens, handwriting is an important clue as to who wrote on a sheet, especially when information is in more than one hand. The type of ink and paper can also speak volumes to those in the know, as can the plant itself. In some cases, it is more elegantly arranged, with the label always in the same place on the sheet. Some people's specimens have more generous proportions than do others and may or may not have more ample information as well. Sometimes a collector will add a witty comment, as this on a golden aster specimen: "*Pityopsis graminifolia:* plants silvery, offering a vaguely cheerful aspect to an otherwise sad landscape, weedy and pathetic with hogweed and rabbit-tobacco, other scratchy things" (fig. I.2).[7] Such information can be recorded instead in field notebooks, again a matter of style, as are collecting habits. Everyone has prejudices, favoring different habitats, ease of access, or in some cases the opposite: the challenge of finding plants in marshes or ravines. There are also outside influences on the collector: patrons, governments, grants.

These are the reasons each herbarium is unique, and why curation is so important. Only intimate knowledge, usually gained over years of experience, can make the most of a collection in terms of communicating its value to others. The influence of the curator is stamped on each collection, private or institutional. An example of the latter is Arthur Tucker, cofounder of the herbarium at Delaware State University, whose enthusiasm shows through in the special collections he developed with the aid of professional and amateur botanists. There is a xylarium of wood samples, a corticarium with a thousand bark specimens, and a fungarium of mushrooms and slime molds. The Ruth Smith Botanical Bead Collection is named for the woman who donated necklaces, bracelets, and other jewelry made primarily from seeds and nuts. There is also an assemblage of potpourri samples, which Tucker analyzed in attempting to identify the species present. Photographs

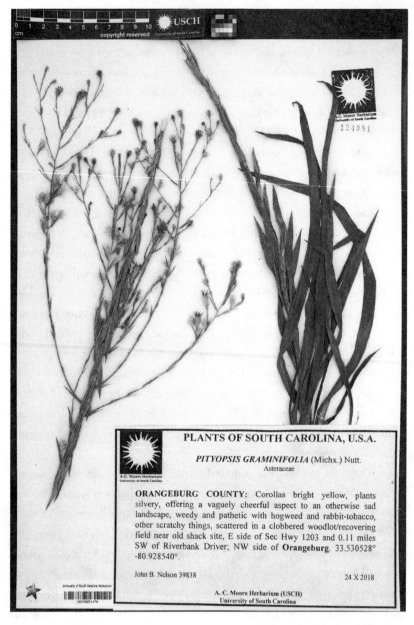

PLANTS OF SOUTH CAROLINA, U.S.A.

PITYOPSIS GRAMINIFOLIA (Michx.) Nutt.
Asteraceae

ORANGEBURG COUNTY: Corollas bright yellow, plants silvery, offering a vaguely cheerful aspect to an otherwise sad landscape, weedy and pathetic with hogweed and rabbit-tobacco, other scratchy things, scattered in a clobbered woodlot/recovering field near old shack site, E side of Sec Hwy 1203 and 0.11 miles SW of Riverbank Driver; NW side of **Orangeburg**. 33.530528° -80.928540°.

John B. Nelson 39838   24 X 2018

A. C. Moore Herbarium (USCH)
University of South Carolina

Figure I.2. Specimen of narrowleaf silkgrass (*Pityopsis graminifolia*) collected by John Nelson in Orangeburg, South Carolina. This illustrates the information that usually appears on specimens today: specimen name with its family underneath; the collector's name; number, location, and date; a brief description of the plant and its habitat; geographic coordinates. The stamp of the A. C. Moore Herbarium and the accession number are in the upper righthand corner. The barcode at the bottom indicates that the specimen has been imaged and the label information digitized; the ruler and color chart at the top are references for quality control. (A. C. Moore Herbarium, University of South Carolina, Columbia)

of botanists who contributed to the collection hang on the walls, and even the bathrooms are decorated with botanical memorabilia. The entire enterprise is a model for a scientifically rigorous institution so welcoming that it would fascinate any visitor. This vibrancy is what many in the herbarium world are now striving for, both within their physical spaces and digitally. As Arthur Tucker and many others envisioned, herbaria are wondrous places, where the plant world can be documented, explored, valued, and preserved for the future.

My goal here is travel through the past, present, and future of the herbarium. I want to look at both the private and public sides of science. Gerald Holton makes this distinction to explain why many people have a rather distorted view of science as being rational, logical, and orderly. This description is true of published reports, which must be understandable and convincing. However, science is actually *done* very differently; it's often messy, difficult, and illogical. Much research begins with a conclusion, a brilliant idea, a hunch for which there is little or no evidence. The process of making sense of an idea and corroborating it may follow a long and winding road, with U-turns, detours, and roadblocks along the way. It is this part of science that holds intriguing stories, including those attached to specimen sheets. The herbarium amounts to a private scientific space, one that includes the intuitive and aesthetic aspects of science. Herbaria are little known outside of botany and taken for granted within the field. The word is not even included in the indices of a leading botany textbook or in another on plant collecting.[8]

The published description of a new species is usually a straightforward report on its characteristics and its relationships to other species. What does not come through is how exciting it was to discover the plant and how beautiful and intriguing its traits are. There is growing evidence that cognitive and affective aspects of an experience are inextricably linked, so it makes sense that science would be emotionally as well as intellectually engaging. As John Dewey argues, any activity has an aesthetic component if one is fully engaged. Consider what it is like to be out in the field searching for interesting plants; the mind and the eye are involved. When a promising plant is found, there are decisions to be made: is this something so rare that it should not be tampered with, is it in bloom, what portion would make a good specimen, how should it be arranged between sheets of newspaper so that it presses well? Then there is recording all the necessary information. The process entails mind, eye, and hand—and, Dewey would contend, also the heart. A recent study shows that collectors tend to select larger, more colorful plants, perhaps unconsciously using aesthetic cues.[9]

The Cartesian separation of mind and body remains ingrained in our collective unconscious, especially when it comes to attitudes toward the difference between art and science: art is about feeling, science about rational thought. Yet both are explorations of the unknown and full of uncertainty; they are complex blends of brain and body activities that are strikingly similar in the two realms. Many botanists have a close connection to art, at least sketching plant structures. Herbarium sheets may have drawings added to highlight fine points of plant form. In some cases, botanists created publishable images, and a few even did their own engravings or lithographs to ensure accuracy. This was particularly true in the age before photography. As will become clear, botanical art has always been important in communicating information about plants. Because of this and the aesthetic aspects of scientific inquiry, art weaves its way throughout this book.[10]

The aesthetics of herbaria also relates to why they need to be preserved and developed in the future. The biologist Edward O. Wilson popularized the term *biodiversity* to highlight the richness of species and their complex interactions, which are undermined with species loss. He also wrote of *biophilia*, human beings' innate attraction to nature. He argued that encouraging this attraction can foster greater respect for living things and a desire to preserve them and the environments where they flourish. In the face of cataclysmic environmental degradation, from habitat loss to climate change, with precipitous declines in everything from frogs to orchids, it would seem that love of nature can have little effect on reversing these trends. Yet Aldo Leopold in his classic *A Sand County Almanac* writes of a conservation aesthetic, contending that the desire to be in touch with nature can take as its ultimate response an impulse toward stewardship, toward protecting and sustaining the natural world.[11]

This is also the argument recently put forth by Michael McCarthy, who sees the usual practical arguments for conservation—the development of a sustainable economy and the economic benefits of ecosystem services— as failing to stem the tide of environmental destruction over the past thirty years. He proposes a different approach, an aesthetically based one, focused on joy and wonder. He frames it as counter to the movement away from beauty in twentieth-century culture that was a response to the ugliness of industrial development and the devastation of wars. Art was overwhelmed by trends toward abstraction and the conceptual, and beauty almost frowned upon as trivial at best. This may also have played into a growing disinterest in plants, the malady referred to as plant blindness that is now being countered by the idea of moving "plants to the foreground of people's hearts and minds."[12]

Millions of people still derive joy from being in touch with nature and its beauty, as evidenced by the greater connection with the living world during the COVID-19 pandemic. If this joy were valued for what it is—an important element of the human experience—then perhaps headway might be made in preserving the rich biodiversity still existing on the planet. Plants are a great source of pleasure. So are collections of dried plants, which might on the face of it seem less appealing, but their stories and the stories of those who care about them are well worth telling. This takes nothing away from the public science of exploring and preserving biodiversity, also covered in this book.

I will explore many themes—specimens and illustrations, travel and discovery, botanical community, the importance of gardening to botanical research, and the significant but not obvious role of women in botany—all of which will recur and morph in many ways. But at the heart of it all are the plants themselves, and the way to best preserve them remains the same method used in the 1500s: pressing them between sheets of paper. The plants are dead, but so much lively, passionate work went into creating the sheets and studying them that they bristle with life.

Rooted in an Herbarium

IN AUGUST 1922, THE AMERICAN BOTANIST Oakes Ames and his wife Blanche arrived at the Berlin train station. They were met by a man Ames knew only through letters, fellow orchid expert Rudolf Schlechter, whom Ames recognized immediately from the orchid Schlechter carried. Oakes spent the next days in Schlechter's lab examining live plants and pressed specimens. Preserved plants, millions of them, were kept in the herbarium at the Berlin-Dahlem Botanical Garden, at the time one of the largest in the world and home to Schlechter's orchid specimens. He also grew orchids in a greenhouse at the garden; the availability of both preserved and living material supported his research in describing species and their relationships. As Oakes worked in the lab with Schlechter, Blanche, an accomplished artist, created watercolor sketches of live orchids that were then pressed and brought home to Boston, where Oakes taught and did research at Harvard University.

A specimen dating from this trip (fig. 1.1) illustrates how much is lost in color and form when plants are dried. Attached to standard-sized sheets (11.5 by 16.5 inches), they are in time usually reduced to shades of brown, no matter how rich their colors were in life. Still, each specimen retains its essential features in a way no drawing can duplicate and serves as a source of DNA and other chemicals for future testing. Since most of the Berlin herbarium collection was destroyed during World War II, this sheet is especially significant, and it is also an example of how image, specimen, and text interrelate—with each element less significant in isolation.[1]

Figure 1.1. Specimen of the orchid *Malaxis dentata* with watercolor by Blanche Ames done from the living plant when it was given to Oakes Ames by Rudolf Schlechter in Berlin; the color bar at the left is a reference for quality control. (Orchid Herbarium of Oakes Ames, Harvard University)

I am presenting Oakes Ames's career because his work exemplifies themes I explore in this book from both the public and private perspectives of botany. It is a little sliver of a larger history. Other botanists may have produced more broadly significant work in the twentieth century, but Oakes's interests were so wide ranging that they point to many aspects of botany, its applications, and how it can be communicated. In addition, the orchid family is one of the most diverse, attractive, and intriguing in the plant world.

By the time he became an undergraduate at Harvard, Oakes was already determined to study orchids. He came from a wealthy New England family, heirs to a railroad fortune. His father developed a fascination with plants and especially orchids late in life, and Oakes became intrigued as well. This was not an unusual pastime in the nineteenth century, though only the wealthy usually cultivated orchids. They were a late nineteenth-century status symbol because they were beautiful, exotic, and difficult to grow without a greenhouse.

After completing his master's degree and setting up a laboratory in his family's home, Oakes studied orchids and published on them. He enlarged his father's orchid collection, expanding the greenhouses to accommodate the plants. Eventually he donated the entire collection to the New York Botanical Garden and concentrated instead on growing his herbarium, revealing one approach to studying plants: relying on preserved specimens. In Oakes's case, they were not all pressed plants. He also had several thousand orchid blooms preserved in jars of alcohol to retain their form and some of their color, as well as a large collection of microscope slides of dissected flowers.

In 1900, Oakes married Blanche Ames (no relation). She had a degree from Smith College, where she had studied art. Her passion was for portraiture, but she spent much of her artistic time creating portraits of orchids for Oakes's hundreds of articles and seven-volume study of orchids, as well as illustrations for the botany courses he taught. She is an example of the many women, frequently relatives of botanists, who contributed substantially to botanical publications, often without recognition, remaining part of the private side of science. Blanche, however, signed every one of her drawings, even those on herbarium sheets. She was a vocal suffragette, and from their letters it is clear that she and Oakes were equal partners in their forty-nine-year marriage.[2]

The basis of Oakes's botanical life was systematics, which includes taxonomy, naming and describing organisms, as well as discovering relationships among them. The herbarium was central to his practice, and his pressed specimens indicate how he used them in his work. Oakes's sheets are noteworthy in that they often included sketches in pencil or ink of portions of the plant, usually the flower, which is often the focus of classification because it has the most specific characteristics. The sketches might be done by Oakes, Blanche, or an assistant and, as with any collection available to other botanists, later researchers also added contributions. A specimen not only documents a plant but communicates about it. Each new annotation adds to its value as a scientific and historical document. The sheets with Blanche's drawings have artistic value as well.

As he began to publish on orchids, Oakes gained credibility, receiving invitations to contribute to orchid treatments in several botanical compendia and to collaborate on a study of Philippine orchids. For this work, he first went to Europe, a seemingly odd detour, but the greatest variety of species from around the world was preserved in the herbaria of colonial powers. Eventually Oakes did collect in Asia, as well as in Latin America, since orchids are more diverse in the tropics. He followed a long line of plant hunters and explorers dating back to the early sixteenth century—many of whose exploits and discoveries were much more dramatic than his, as will be described in later chapters. Oakes was foremost a systematist, working on describing orchids rather than collecting them, so he relied on others for many specimens, paying for some collections and trading specimens he had in multiples for ones he needed, both common practices in the herbarium world. He also benefited from the tradition in botany of being entitled to keep a specimen of a species he identified for another botanist. Borrowing from other collections or visiting them were also important means of communication among botanists, helping to fill information gaps. In all, Oakes published on over a thousand new species: some he discovered in the field, others were sent to him, and a number were revised identifications. Taxonomy is a continuing process of refining species definitions as more information becomes available.

After he and Blanche had lived for six years in his mother's home, where Oakes had set up his laboratory and herbarium, Blanche convinced him to move. A crisis contributing to this decision developed when the children's nurse came down with pneumonia, a frightening infection in the pre-penicillin era. To keep the children safe in case others in the house became ill, Blanche decided to take them to her parents' home. Oakes did not take kindly to this, believing that his wife should have gone to stay with his brother, who lived closer. In a letter he tells her that her fears "had no foundation in fact," and accused her of not being able to see her way clear to "remain under a few inconveniences." Then he raises the real problem: he insists that she must finish the drawings for his book and not "fritter away time with gossip." Not surprisingly, Blanche, suffragette and what her husband called a "new woman," responded that when she tried to discuss the problem with him, "you did not take the trouble to put down your herbarium sheet and your [magnifying] glass, but with one eye screwed up and the other on a dried flower, you answered me in scarcely more than monosyllables. . . . A few moments in the herbarium showed me that I could expect no aid."[3] This is the private side of science at its most personally intense.

On farmland nearby, the Ames family built a house that included a two-story library and workspace for Oakes and his collaborators. Meanwhile, like many botanists, he became involved in more than just systematics. He taught botany at Harvard, where he eventually took on administrative roles. Almost to his surprise, he turned out to be an able manager, and he and Blanche bought a second house in Boston so he could be closer to his work. Oakes moved his herbarium and library there, eventually donating both to Harvard in 1939. Thus his private collection of over sixty-four thousand specimens, eighteen thousand microscope slides, twenty-five hundred flowers in alcohol, and thousands of books became an institutional collection, as happened with many herbaria as botany became more organized over the centuries.[4] Also included were all of Blanche's sketches and illustrations. So the botanist's three essential tools—specimens, texts, and images—remained together and augmented one another's value. Just as Blanche's drawings often documented orchid structures more effectively than photographs could, the published literature was essential to Oakes's work because botany is a historical science in that botanists must compare their claims about what counts as a species to what earlier botanists have described.

In 1923 Oakes assumed directorship of the Harvard Botanical Museum from George Goodale, who had founded it in 1886. Such collections were popular in the late nineteenth and early twentieth centuries, when there was great interest in economic botany, the study of how people use plants. Its subjects included food plants and their processing as well as medicines, textiles, paper, household items derived from plants and, of course, timber. Several years before Oakes became museum director, Goodale encouraged him to teach a course in economic botany. Oakes plunged in with enthusiasm, wrote a book on economically important annual plants, including grasses, and had Blanche create large posters, including evolutionary trees showing the relationships among economically important species.

As discussed in the next chapter, useful plants, particularly those with medicinal properties, have been central to botanical inquiry since ancient times because they matter so much to people, and that interest continues to this day. In Oakes's book he describes species that provide the world with drugs such as cannabis, the opium poppy, and tobacco, and familiar foods like peas, corn, gourds, and grains; he also traces the history of their domestication and use. One of his students, Richard Evans Schultes, who eventually taught the economic botany course himself, was intrigued by psychoactive drugs. He spent a lifetime studying their use by indigenous peoples in Latin America, contributing to what has become the field of ethnobotany. Along with his work on classification and his collection of thousands of specimens, Schultes also learned about the cultural practices surrounding these plants and recorded the deep knowledge of those using them. He was both respectful of traditions and mindful of the religious significance as well as the usefulness of these species.[5] He and Oakes also raised alarm about the encroachment of development on formerly isolated forest peoples and the related loss of biodiversity.

Besides teaching, Oakes's roles at Harvard included directorship of the university's botanic garden and later of its Arnold Arboretum, an extensive collection of living trees. Its first director, Charles Sprague Sargent, had not only developed the grounds but sent a number of plant collectors to

China and other areas in search of new species that might be useful in horticulture, agriculture, and forestry. Sargent built a sizeable herbarium with these collections, and Oakes named a retired plant collector, Ernest Wilson, as its keeper. From his studies in economic botany, Oakes was well aware of botany's ties to gardening and farming. He maintained the arboretum's collecting and research activities in studying tree varieties that might have horticultural value. He was ever mindful of the importance of herbaria in documenting such activities.

One aspect of herbaria often taken for granted is the technology related to the preservation and study of plants. In one sense, herbarium practices are much as they were at their inception in the sixteenth century: press plants between pieces of paper and boards; the paper absorbs moisture and the boards keep the plants flat. Then the dried specimens are pasted to paper, labeled, and stored. However, over the centuries, metal cases, which are more pest proof, have replaced wooden cabinets, and labels are printed instead of handwritten. In addition, Oakes was a proponent of adding photographs to specimen sheets; he was particularly interested in photographing type specimens when he visited Europe. A type specimen is one studied for the first published description of the plant and therefore is especially important in verifying a species' identity. At the time, the greatest number of types were in European herbaria. Photographs were a way for Oakes to bring home the basic information on a sheet. This example of his working method shows how the private side of science reveals itself as he examines type specimens: "One of the thrills of my career came in Paris when I turned with breathless interest to . . . types and drawings to see at last just what was meant by hopelessly obscure words. And then to pin up these precious relics and photograph them in the dim light which filters through dusty window glass, glass which has probably not been cleaned since Napoleon's time. You're in a sense of happiness I shall not attempt to describe. Once a systematist becomes a slave of types, his contempt for guesswork reaches dizzy heights. Never again can he become content with the uncertainty of words and identification by supposition. Surely the unrest in my soul, caused by doubt, made me determined to represent in my herbaria by every possible means the types of orchids."[6]

Studying a photograph was not a substitute for examining the plant itself, but it did augment written descriptions, providing additional information and, as Oakes described, verifying it. A photo was a way to record visual evidence of specimens that were relatively inaccessible to many botanists without the expense of extensive travel. A number of other botanists during the first half of the twentieth century took up this practice, which was a harbinger of the specimen imaging and digitization of label information now being done on a massive scale around the world. However, access remains a significant issue as developing nations grapple with conserving biodiversity and attempt to build botanical knowledge infrastructures locally; the preponderance of type specimens and taxonomic information is still in the developed world.

Oakes also made good use of printing technology, drawing on his own resources and Harvard's to create a printing operation in the Harvard Botanical Museum, where his volumes on orchid taxonomy as well as periodicals and other books were published. Many botanical gardens and museums follow a similar model to communicate both to other botanists and to the larger community of plant lovers. Plants' beauty attracts people to them, and obviously Oakes was aware of this, particularly because of his collaboration with Blanche. She mastered the art of creating scientific illustrations in pen and ink and the conventions of enlarging small structures to clearly delineate subtle characteristics with finesse and style, as do all good botanical artists. Blanche also created etchings of orchids, and Oakes encouraged her to publish a book of her orchid illustrations, for which he wrote the text, introducing Blanche's art to an audience beyond the botanical community, as did their work with the Orchid Society of America.[7]

Another way that Oakes developed the link between botany and art was through continuing a long-time collaboration that had been begun by Goodale in the 1880s when he contracted with the German glass artists Leopold and Rudolf Blaschka, father and son, to create realistic glass sculptures of plants for the Botanical Museum. This work continued until Rudolf died in 1930. Funded by Elizabeth Ware and her daughter Mary, who had been Goodale's student, the collection has become world famous for its beauty, fragility, and botanical accuracy. To highlight it, Oakes wrote a guide to the collection that became a best seller at the museum.[8] He was aware of the importance of beauty in attracting people to learning about plants, but the Blaschka plants were meant for study as well. Just as preserved specimens would be, they were attached to standard-sized herbarium sheets as a reminder that the models were meant to be more about science than art. They were also arranged according to the latest classification system. They exemplify using art to communicate about plants, a practice that goes back to the time of the first herbaria in the sixteenth century, sometimes attributed to the Italian botanist Luca Ghini, who often sent sketches or prints along with specimens to his colleagues.

In the scope of his interests and his access to financial resources, Oakes Ames was hardly a typical twentieth-century botanist. He was more reminiscent of those in prior centuries when botany was not yet a profession but an avocation, albeit often a serious one. Through his story are woven the major themes I address in this book. The scientific study of plants is at the core of why herbaria exist and why they are valued, but the economic and cultural significance of plants have always been major drivers of that science, especially of the collecting explorations begun in the sixteenth century. Medicine, agriculture, and horticulture were on the minds of those who sought new species, motivated by desire for financial gain, status, or entry into wider social and scientific circles. Building collections was an obsession even among some of the earliest herbarium creators, though the importance of this practice became more appreciated over time.

Early Botany

IT'S OFTEN DIFFICULT TO PINPOINT where and when a good idea first arises. The Italian botanist Luca Ghini is credited by many with creating the first herbarium, but the picture is not totally clear. His herbarium no longer exists, but the latest research suggests he was collecting plants in the 1540s. Scattered examples of pressed plants exist from the late fifteenth century, but these are not really collections, little more than a few leaves pasted into notebooks.[1] A variety of origins would not be surprising since pressing leaves or flowers between pages of a book seems an obvious way to preserve them, as scientific or romantic remembrance, though at the time books were not that common. Most agree that Ghini created an herbarium with hundreds of pressed specimens and also spread the practice among his colleagues and students. But he did much more than that as a pivotal figure in promoting practices crucial to the development of modern botanical inquiry.

Ghini emphasized direct observation of nature as the surest path to accurate knowledge. He was in the second generation of early modern scholars groping toward new ways of learning, knowing, and seeing. At the University of Bologna, he was schooled by medical faculty who taught the importance of direct observation in medicine. Yet Ghini was also well grounded in the intellectual traditions of the past, including reliance on ancient texts. In medical education, essential authors included Aristotle's student Theophrastus, who presented an organized description of plants, and the Greek physician Dioscorides on *materia medica,* the study of medicinal substances, many derived from plants.[2] Ghini knew these writers and relied on them, but he was convinced that texts were not enough, that physicians had to learn by seeing plants for themselves.

As Ghini progressed in the medical profession, teaching at the University of Bologna, he was drawn more and more to studying plants used in treating disease. He devised ways to observe closely, document what he discovered, and share that knowledge with students. He took notes on the plants he found growing in the wild and in gardens. He sometimes added drawings along with pressed specimens to explicate a structure or the growth habit of the plant. He would cut out and include printed images, which became more available in the 1530s with publication of the first printed herbal, or book on medicinal plants, which included accurate illustrations drawn from life. The text was by Otto Brunfels and the images were those of Hans Weiditz. In 1542, Leonhart Fuchs and a team of artists produced another great herbal of the period, revealing to botanists like Ghini the possibilities of communicating information with accurate images.[3]

The most direct way to document a plant's characteristics was to present the plant itself, either growing or preserved by simply pressing it between two pieces of paper and flattening it between the pages of a book or wooden boards—anything to keep it flat. Once dried, the specimen was relatively indestructible. Its tissue would not rot, and keeping it flat meant it did not curl into an unrecognizable twisted mass, as happens after a frost. The term *herbarium* for a collection of such plants was not introduced until the late seventeenth century. Earlier, "herbarium" referred instead to written works on plants, either manuscripts or printed books, often with illustrations, while a collection of pressed plants was called a *hortus siccus,* the Latin for dry garden, or *hortus hyemalis,* winter garden, particularly useful during winter months when there was little plant material to examine outdoors.[4]

While physicians might prescribe plant-based medicines, they had often left plant collecting and medicinal preparation to apothecaries, who also created herbaria and began to more seriously study plants. Increasingly, physicians began to study medicines more closely; Ghini's penchant for direct experience of nature meant that leaving this work to apothecaries was no longer good enough. Along with collecting and preserving specimens, Ghini also experimented with growing plants and testing their medicinal properties. He became so involved in plant studies that he created the first course in materia medica for future physicians. The University of Bologna allowed him to take on this extra work, although with no change in title or salary. Eventually, he was named professor of materia medica, but his working conditions were still not good, so in 1543, when Cosimo I de' Medici, the grand duke of Tuscany, invited him to move to Pisa and teach at the university there, Ghini took the offer.[5]

Medici sweetened the deal by asking Ghini to establish a botanic garden, the first of its kind for studying and teaching about medicinal plants; earlier gardens were used more as sources of medicines than for study. Medici himself avidly explored botany and had his own palatial gardens. The early modern period saw a renewed interest in growing plants, not only as food and medicine but simply for the pleasure of seeing them, walking among them, and showing off what could be grown given enough money and horticultural expertise. Medici and his sons were also interested in the medicinal uses of plants and had their own laboratory for preparing medications.[6] Some recipes were quite elaborate, and there were competing versions in the various translations of ancient works like those of Dioscorides, so there was need for direct experience to gain accurate knowledge.

In Pisa, Ghini would teach and then take students into the garden to examine the plants he had discussed. The herbarium was a useful adjunct even when the weather was good; it was helpful to see the same plants alive and dried so differences would be apparent, making it easier to later identify the pressed plants. The link between herbaria and instruction is strong throughout botanical history. Students were, and still are, often required to make their own collections to help them appreciate the value of reference material. Several of Ghini's students and colleagues created herbaria, some of which still exist. One begun in 1558 is attributed to Ghini's colleague Francesco Petrollini; the sheets are now bound like the pages of a book (fig. 2.1).[7] This format probably aided the collection's survival. Shuffling one page against another can loosen specimens or rub off bits of the brittle plants. Binding reduces the chance of damage, but also makes the collection less flexible. As more and more plants were studied and as herbaria grew,

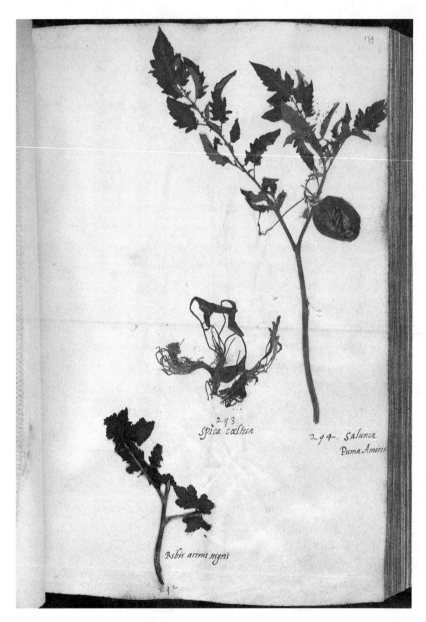

Figure 2.1. At the right, a fruiting specimen of tomato (*Solanum lycopersicum*) from the En Tibi herbarium, now attributed to Francesco Petrollini of Bologna and dating from about 1558. (Naturalis Biodiversity Center, Leiden)

the need increased to organize the specimens according to similarities among them, for example, putting all the grasses or ferns together. This was difficult with bound volumes. Also, loose specimens could be lent more readily to colleagues.

A Communal Activity

From Ghini's time on, plant collecting and study became increasingly communal activities. He developed the practice of joining students and colleagues for field trips to study plants in situ and to collect specimens. Learning wasn't just about taking notes; it involved seeing plants growing, handling them, and identifying their characteristics. Some excursions were in the vicinity of the university, but Ghini also went further afield, including a trip to the island of Elba and another to Monte Baldo in the Italian Alps; different geographic areas meant access to different plant species. Others traveled to see Ghini. Valerius Cordus visited from the Netherlands, William Turner from England, and Guillaume Rondelet from France; all three were influential members of their countries' botanical circles. Networking drove the rise of early modern botany; Turner picked up the herbarium idea from Ghini and Rondelet may have gotten Ghini interested in field trips. The Swiss botanist Conrad Gessner claimed that field trips provided the purest form of earthly pleasure: to do as well as see, to move through the landscape and discover, and to share these encounters with colleagues and students. But field trips were evanescent activities; that's why notes, drawings, gardens, and dried specimens were needed as reminders of them— memories turned into knowledge.[8]

Distributing this information became the heart of the botanical enterprise. Ghini was intellectually generous and shared his specimens not only with his former students but with others passionately studying plants. The German botanist and herbal author Leonhart Fuchs traded specimens, illustrations, and notes with Ghini. After founding the Pisan garden, Ghini organized seed collection at the end of each growing season and then sent lists of seeds available to others. Germinating seeds resulted in more plants that could be studied and made into specimens. These elements of plant study—seeds, specimens, notes, images—are what one sociologist calls "immutable mobiles," items that can be easily moved and still communicate the same information, making them powerful tools in research, in building research communities, and in allowing facts to travel.[9] Since this was long before the age of postal systems, correspondents relied on traveling friends to pass on items or pick up parcels. These favors were usually reciprocated with specimens, useful information, or introductions to others who might be helpful.

As in most networks, there were key nodes, particularly among well-connected individuals. Ghini was one; another of the next generation of botanists was also from Bologna, Ulisse Aldrovandi. He created an herbarium and described a paste to hold down the specimens on paper. Aldrovandi spent time in Pisa learning from Ghini, who shared research and teaching materials that were helpful when Aldrovandi became professor of natural sciences at the University of Bologna. Together they investigated the plants in the Pisa botanical garden, and the younger man published a catalogue of 620 species growing there.[10] Such lists, containing brief plant descriptions, became a common means of botanical communication and of advertising a garden's noteworthy species.

Aldrovandi was driven by a passion for learning about the natural world and saw possessing nature as a way to understand it. At the time, natural history was not a segmented discipline as it is today. Spending time in nature meant exposure not only to plants but to wondrous birds, insects, and other animals, so collections were rarely restricted to one group of organisms, nor were the accompanying observations, which might include comments on animals, landscapes, and rock formations encountered on field trips. Aldrovandi had the private means to create a massive collection as well as a large library with volumes of observational notes and illustrations he commissioned from notable artists to document the collection. He thus fashioned one of the first museums, distinguished from the more common cabinets of curiosities of the time by being organized for study as well as for display.[11]

The huge size of Aldrovandi's collections, with thousands of herbarium specimens alone, suggests that accumulation itself became a passion; collectors live through their assemblages. In his case, the zeal related not just to scientific curiosity, though that was a major driver, but also to the prestige that came with a large, well-presented collection that spurred natural history investigation. Since comparison was at the heart of the botanical inquiry emerging at the time, more material available meant better comparisons could be drawn: newly encountered species related to similar ones already studied.

In 1554, Aldrovandi was joined by Ghini, who returned to Bologna from Pisa and took up his old position at the university, though he was in poor health and had come home to die. Aldrovandi assisted him in his work and helped him sort his collections and papers. Because of his liberal sharing, many of Ghini's specimens and notes were on loan to other botanists who were working on their own botanical investigations. While Aldrovandi was seen as his heir apparent, with the expertise to fill that position, the person who benefited most from Ghini's knowledge and resources was Pietro Andrea Mattioli. Ghini is hardly remembered today, and Aldrovandi is recalled mainly because a portion of his museum still exists in Bologna, but Mattioli's publications made him the best known of the three.

Ghini's Protégés: Mattioli and Cesalpino

Mattioli graduated from the University of Padua's medical school. Unlike Aldrovandi and Ghini, he did not go into teaching but practiced medicine, eventually moving to Vienna and becoming the Holy Roman Emperor Maximillian II's personal physician. Mattioli's thirst for plant knowledge focused on better medical treatments, and like many in this era he considered the best way forward was through revising the work of Dioscorides, a Greek who lived and worked in southern Turkey in the first century CE. In 1544, Mattioli published an Italian translation to make the text available to those who did not read Latin, as the use of the vernacular was becoming more popular in printed works. He included commentaries correcting or expanding on the original text, a common practice at the time.[12] However, when Mattioli was writing, there were forces at work making Dioscorides less reliable: botanists were coming to realize that different plants grew in different regions.

This presented a problem for physicians and apothecaries: how could they know that a plant Dioscorides associated with certain medicinal properties was the plant they had at hand, since the species he described were from the eastern Mediterranean? Some saw this as a philological issue, a matter of textual descriptions being translated from Greek to Latin and then into vernacular languages like Italian, French, German, and English. They attempted to work out difficulties by editing Dioscorides and adding commentaries, as Mattioli did. Others, and these became more influential as the sixteenth century progressed, sought the solution in direct observation of the plants they had before them and testing the species' medicinal properties. Ghini tended to use the latter approaches, and Mattioli was the recipient of his research.

Around 1551, Ghini made a conscious decision not to publish any work, although he had notes and illustrations for a projected natural history. While he appreciated the importance of illustrations, they were expensive to publish, and he was not in a position to take the necessary financial risk. Instead he shared his work with Mattioli, who had not studied with Ghini but sought his advice because of his reputation. Ghini wrote annotations to Mattioli's commentary on Dioscorides. Called the *Placiti*, it was composed of sixty-nine opinions or notes on plants. He sent it to Mattioli, recommending that illustrations be included. Ghini also shared drawings and over a thousand pressed specimens, but Mattioli had a different attitude toward preserved plants than did his elder. Although Ghini kept an herbarium, Mattioli threw away his dried plants and even his drawings after studying them.[13] Such a cavalier attitude became less common as time went on, and Mattioli later regretted his shortsightedness. In the mid-sixteenth century the herbarium and botanical methodology in general were in development, and the specimen's place in inquiry was not yet fully appreciated. It was a time of experimentation in many ways, including what counted as reliable information.

In 1554 Mattioli published a Latin translation of Dioscorides that incorporated Ghini's notes. It was illustrated; later editions had even more woodcuts, making Mattioli's work of lasting interest to botanists and book collectors.[14] The text was still tied to ancient knowledge even as other writers were producing works that relied more heavily on direct experience and observation. Experimentation by naturalists like Ghini was emerging at least in rudimentary forms, such as sowing seeds, observing plant development, figuring out the best growth conditions, taking notes on the results, and then using the information to improve conditions for growing that species or others. The basic process was something gardeners and farmers had done for millennia. What changed was that information was now recorded and formally shared with others; this knowledge soon became more valued than ancient texts. There was also a move away from emphasis on medicinal uses of plants to studying a broader range of species for their own sake, for their myriad forms and structures, and the varied habitats where they grew.

One of Ghini's former students and colleagues took this a step further. After finishing his education, Andrea Cesalpino remained in Pisa, working with Ghini. When Ghini left Pisa in 1555, Cesalpino replaced him as professor and director of the botanical garden. He also adopted his mentor's device for documenting plants for teaching and research: the herbarium. He created at least two, one for study and teaching and another, with over seven hundred specimens, that he dedicated to a local bishop. Herbaria were becoming what those in science studies call "boundary objects," tools employed for a variety of purposes.[15] While Ghini saw teaching and documenting species information as the herbarium's foremost purposes, Cesalpino also considered the form a way to make his learning better known and to thank a patron.

Cesalpino wrote about the plants he studied in the field, the botanical garden, the herbarium, and the literature. Beyond collecting and organizing information, he sought fundamental concepts behind the details. Despite being a physician, Cesalpino did not, like Mattioli, focus on the medicinal properties of plants. He took a more theoretical view of natural history. Like Aristotle's pupil Theophrastus, he attempted to formulate a plant classification system, identifying basic traits to develop his taxonomy. Theophrastus divided plants into four categories—trees, bushes, shrubs, and herbs—noting that the division among them is not hard and fast. This is a deep classificatory problem that plagues systematists to this day: living things refuse to fit into neat categories.

Cesalpino did not have his specimen sheets bound into volumes; he kept them loose so he could compare species, reorganizing the pages to refine his classification method. He put trees and shrubs together and had a second category for seeds without coverings—the gymnosperms, including pines, cypresses, and firs. Then came herbs—non-woody plants with covered seeds—and finally plants without discernable seeds, such as mosses. Cesalpino was one of the only early modern botanists to conceptualize an underlying order among plants. This issue later became urgent: more and more species were discovered as botanical field trips made the flora of Europe better known and worldwide explorations brought in a bounty of new species. Even in the sixteenth century, plants from the Americas were being grown in Pisa, with Cesalpino discussing the tomato, sunflower, and agave. He understood the way botany was heading with more species to investigate and felt it necessary to devise the basic principles of plant organization.[16]

Conrad Gessner

Italy was hardly the only area where plants were studied intently. Medical education in northern Europe was also focusing more on materia medica, with Guillaume Rondelet in France and Valerius Cordus in Germany in the fore. They had both studied in Italy and met Ghini; they then returned home to educate other botanist-physicians. Many others from northern Europe also traveled to Italy to avail themselves of the latest in medical education. They learned to observe and record carefully and to look at plants for their own sake as well as for their medicinal properties. They used herbaria as recording devices and memory aids for what they had seen in the south, thus making collections more widespread.

As Ghini was a central figure in the Italian botanical network, the Swiss naturalist and physician Conrad Gessner was important in the north in a variety of fields. He produced dozens of books, in part to support his large family, but his drive also arose from a humanist passion to know more about the world. This fueled several different approaches to gathering knowledge. Gessner kept an herbarium and also collected zoological specimens. He was an avid note taker and sketched as well in order to organize his growing store of information. Gessner's wide circle of correspondents provided him not only with informative letters but also the loan of books, specimens, and drawings. Books were expensive and many were rare, so ownership was limited, but the information they contained was invaluable to scholars like Gessner who wrote compilations on a variety of subjects, from animals and plants to philology and geology. To repay those who were generous in sharing their resources and knowledge, Gessner was equally generous in thanking them in his publications, even naming species for them. His letters often held out the carrot of acknowledgment as a reward for sharing.[17] This was to become a common practice among later collectors, especially naming species after those who were particularly helpful.

In the botanical sphere, Gessner, too, tackled Dioscorides in a Latin edition with commentary by Valerius Cordus, who died young of malaria. Gessner also published Cordus's manuscript of plant descriptions, considered important because these were so thorough, giving Gessner exemplars in writing about plants. He produced a list of plant names in Latin, Greek, German, and French as a way of reconciling information published in different languages—very much in the tradition of textual study by early modern naturalists. It was an attempt to organize plant knowledge, resolving names across language barriers so physicians and apothecaries could be more certain that they were all referring to the same plant. It might seem relatively simple to link a plant with its name, but one plant could have many names, used not only in different geographical regions and different languages but in different cultural contexts. Physicians, apothecaries, and laypersons might have various terms for the same plant or, equally confusing, the same term for different species. This knotty problem in some ways persists to the present day. In settling such confusion, specimens have long been essential arbiters.

While he wrote on many topics, for at least the last ten years of his life Gessner was planning a companion to his *Historiae Animalium:* an ambitious illustrated work on a broad spectrum of plants. In preparation he kept voluminous notes, often devoting an entire page to a species. These notes were both textual and visual, and fortunately many of them survive.[18] The drawings were in pen and ink, most at least partially colored. Gessner's notes describe characteristics of the plants, including their habitats, growing habits, and ranges. He included quotations from ancient writers and from correspondents such as Mattioli, Fuchs, and Aldrovandi, with whom he shared specimens, notes, and drawings. His notebooks really were meeting places of many minds, of many perspectives on a plant. Though Gessner's herbarium no longer exists, his correspondence indicates that he had one, borrowing and trading specimens freely.

Gessner's concept of a particular species could change over time as notes and drawings were added and amended. A case in point is a specimen he first encountered in 1554 and could not identify. He made a drawing and kept adding information over the years as he came upon the plant—in life and as a specimen—until finally he discovered it was tobacco, *Nicotiana,* which had been introduced to Europe from the Americas (fig. 2.2). This is a beautiful example of building knowledge of a species over time, and he did this for over eight hundred plants.[19]

Looking through the pages of Gessner's notebooks, there is no doubt that his understanding of plants was profound. As with tobacco, he often returned again and again to add more information on particular species, including further drawings of flowers, fruits, and seeds: in other words, different parts of a plant's life cycle. For many species Gessner also had available to him haptic encounters with living and preserved plants that went beyond drawing. He would handle and physically examine a plant. He would explore its textures and smell and taste it, thus connecting with the plant as fully as possible using these closely related epistemological techniques. This fits John Dewey's definition of aesthetic experience: engaging as Gessner did with all his senses, his keen intelligence, and his artistic hand.[20]

Carolus Clusius

Just as there were many styles in herbaria and in presenting specimens and plant illustrations, there were many approaches to studying them, as the life of the Dutch botanist Carolus Clusius suggests. His career and interests point forward to seventeenth-century botany with his involvement in travel,

Figure 2.2. Conrad Gessner's notebook page on Brazilian tobacco (*Nicotiana rustica*). (University Library Erlangen-Nürnberg, MS 2386, fol. 13 r)

horticulture, and exotic plants. He studied medicine at Montpellier, then a flourishing center for materia medica, in large part because of the inspired teaching of Guillaume Rondelet. Like Rondelet, Clusius had an herbarium, though it no longer exists, yet he did use specimens in his work and actively sought them out to learn more about a species.

Clusius studied medicine but never practiced, pursuing instead his interest in plants. Coming from a noble family, he always managed to find patrons to support him and his botanical work. From 1563 to 1565 he traveled in Spain and Portugal as tutor to a wealthy banker's son. The father wanted Clusius not only to educate his child but also to learn about the exotic plants coming into Spain from the New World and their possible economic value. It becomes clear in this book's later chapters that throughout the history of plant collecting, economic and political aims have been interwoven with scientific ones. Clusius collected specimens and seeds, ultimately wrote a book on Spanish plants, and studied American plants like the potato. He also shared his knowledge and seeds with his next patron, a nobleman in the Netherlands with a passion for gardening, which remained central to Clusius's interests in later years.[21]

Clusius had a gift for languages, mastering eight; this facility allowed him to connect with correspondents throughout Europe, to handle plant nomenclatural issues, and to translate the botanical writings of others, giving them a broader readership. He began by producing a French version of the botanist Rembert Dodoens's significant survey of plants, first published in Flemish. In this and his other translation projects, Clusius followed the common practice of reorganizing the work and added commentary and new information. While in Spain, he learned of Nicolás Monardes's book on Latin American plants and translated it into Latin, making it accessible to naturalists across Europe who were hungry for information about exotic plants. Clusius translated two Spanish works on East Indian plants into Latin in the 1570s, and also created Latin editions of the voyages of two British explorers, Francis Drake and Walter Raleigh. Clusius included descriptions of cocoa and sweet potato grown from Drake's seeds.[22]

Clusius's language abilities were also valuable because the Netherlands was then under the sway of the Hapsburgs and the Holy Roman Empire, which ruled Spain and much of what is now Germany and Austria and also had links with the Ottoman Empire centered in Constantinople. The Dutch used these connections in their shipping ventures, and there were Dutch diplomats holding powerful positions throughout the empire. One of these, Ogier Ghiselin de Busbecq, who had been Hapsburg ambassador to the Ottoman Empire in Constantinople, was a serious gardener, and brought back plant materials from his travels. He supplied Mattioli, then a physician to the Hapsburgs, with specimens and seeds of new species such as the horse chestnut, which was described in the latter's revision of Dioscorides and also propagated by Clusius and his associates. Busbecq supported Clusius's bid to become director of the imperial medical garden in Vienna and also introduced a number of flowering bulbs from the Near East. Plant enthusiasts like Clusius were always on the lookout for new wonders. Clusius served in Vienna from 1573 to 1576 and took the opportunity to make collecting trips into the mountains of Austria and to develop contacts with those traveling in the Near East.[23]

Adding to his influence, Clusius continued writing. He published the first book on fungi, having studied the mushrooms of central Europe during his time there, and followed it with the first book on Austrian and Hungarian alpine flora. Along with his work on Spanish plants, these were significant contributions to botany at a time when European flora had yet to yield many of their secrets. Toward the end of his life, Clusius published a four-volume compilation of plants with woodcut illustrations and four years later an additional work on exotic plants that he had received from his many contacts; it was one of the broadest plant studies of the time.[24]

Clusius's impact grew as he worked to maintain connections with those from whom he could acquire interesting plants and information. Like Ghini and Gessner, he was at the center of a large botanical network. There are thirteen hundred letters to Clusius from over three hundred people in six languages preserved in archives.[25] Their emphasis is definitely on horticulture, on plants for their own sake and for their aesthetic value. His correspondents shared these pursuits and, like him, their passion fired keen observation. They were also literate and intelligent enough to provide useful insights and share seeds, specimens, drawings, and information. Clusius could then pass these on to others or describe them in his publications, so this was hardly a one-way street from expert to client. There were a number of women in his network, and he was particularly successful in nourishing their interest. Most were from the upper classes, with the education, time, and money to develop their gardens. They were observant, passionate, and willing to spend money on plants, books, and illustrations so they could learn botany.

Clusius visited members of a thriving botanical community living in London for at least some portion of their lives. Several were termed the Lime Street naturalists since they lived on or near this street; those studying plants collected specimens and kept herbaria.[26] They are little known today because most were content with trading information and specimens and did not publish, repeating Ghini's fate. Clusius also had correspondents in Italy and contacts in France, including the king's gardener, Jean Robin, who distributed specimens and seeds from North American plants growing in the royal garden.

When Clusius finally returned to the Netherlands in 1593, he became a professor at Leiden University and prefect of its new botanical garden. He brought his network to bear in providing seeds, bulbs, and plants from around the world and worked with an apothecary who kept an herbarium to document what was grown. In part because of the garden, the university grew into one of the most important institutions for medical training in the following centuries. After the Dutch East India Company, a governmental trading organization, was set up in 1602, Clusius wrote out instructions for its personnel on how to collect seeds, cuttings, and specimens and pack them for transport. When the first company ships returned from the East, he traveled to Amsterdam to interview those who provided plant material and information. He was in his seventies by this time, but never lost his thirst for new botanical treasures.

This chapter has focused on botanists like Clusius whose specimens and ideas shaped the development of early modern botany. The next one will examine the tools they used in learning about the plant world, including those needed to preserve specimens and to present their findings in illustrations and texts.

THREE

The Technology and Art of Herbaria

𝒦

AS EARLY MODERN BOTANISTS collected plants and learned about them, they took advantage of innovative technologies and techniques that made possible different forms of botanical inquiry. Botanists were asking new questions about plants, using new tools in answering them, and making their discoveries known to others. Novel ways of seeing and of disseminating information were maturing. Humanist studies were broadening horizons and the arts were achieving greater realism.

One of the tools most used by botanists was paper: to dry and mount specimens, to take notes, to publish their ideas. Without paper, there wouldn't be a specimen, if for no other reason than that it had to be pressed between sheets of paper to absorb moisture. Water allows mold and other agents of rot to have their way; remove the water and plant tissue is much more stable. It is thought that Luca Ghini first kept loose specimens between pieces of paper, but soon animal glue was employed by some to keep plants in place. Alternately, a plant could be sewn down or slits made in the paper so the stem could be slipped through—two techniques that made it easier to remove the plant if its owner wanted to examine it more closely or pass it on without the paper. In any case, the species' name, if known, had to be added to the sheet, sometimes with the place and date of collection. The latter information became more commonly recorded over the centuries.

When the first herbaria were created, paper was an expensive commodity, its manufacture introduced into Spain from the Islamic world in the late eleventh century. Paper was produced in Italy from 1235, but it was not

made in England until 250 years later.[1] Paper only slowly replaced the more
durable parchment in manuscripts, though preparing animal skins as a
writing surface was laborious and expensive. With the invention of movable
type in the mid-fifteenth century, demand for paper increased, leading to
cheaper, more readily available forms.

Paper is essentially an unwoven mat of cellulose fibers derived from
plants. In the East, mulberry fiber was the favored starting material, but in
the West most paper was made from linen, hemp, or cotton rags cut up and
then soaked to loosen the fibers through fermentation. This preprocessed
plant material was reduced to a watery slurry, then poured into a mesh-lined
frame. The water was pressed through the mesh and the damp sheets hung
up to dry, then further flattened and smoothed. Paper was usually sized, that
is, coated with gelatin to make the surface smoother and more waterproof.
With little sizing, paper was good for pressing plants because it readily ab-
sorbed water, but not good for writing because water-soluble ink soaked
into the fibers, resulting in blots. When printing presses were invented, the
paper not only needed to be sized, but the process called for a different ink,
an oil-soluble one thick enough to adhere to the metal type's surface.[2]

Early modern naturalists employed these technologies: paper in
pressing plants and ink in writing observations, plant lists, and letters. They
also wrapped seeds in paper to keep them dry and to pack notes, books,
and specimens to send to other botanists. Several types of paper, varying in
cost, were needed for these purposes. Paper's value is suggested in the recy-
cling of old letters and notes for pressing plants or packaging seeds, and in
pasting specimens to both sides of a sheet, something unheard of today
when care is taken never to turn over an herbarium sheet for fear of loosen-
ing any of the plant material.

Naturalistic Drawings

Drawings of plants were also often done on paper. The pages of Conrad
Gessner's notebooks are filled with drawings and explanatory notes that
made his finds concrete and provided resources for further study. He
pleaded with correspondents to send sketches if they could not send the

plants themselves. This was particularly important for rare plants and those that didn't press well; orchids and succulents, including cacti, often fell into this group. Gessner even offered to pay for having an image drawn, but was careful to mention that he wanted only rare plants, ones he hadn't already recorded. He did many drawings himself and worked with artists, even standing over them to make sure they represented structures correctly, in a collaboration termed "four-eyed sight" that could deepen over time as each learned to see differently through the eyes of the other.[3]

Though they were left to the same person, Gessner's notebooks with their drawings survived but his herbarium did not, suggesting that drawings may have been considered more valuable then specimens. At the time, pressed plants were useful in studying and differentiating among species, but there was little thought of them as documents that could gain value with time. Especially for readily available species, one specimen would have been considered as good as another, so disposing of them would not be given much thought, as was the case with Pietro Andrea Mattioli's plants. Good drawings, on the other hand, were more difficult to come by, and though many of Gessner's were not published until two hundred years after his death, those who acquired them appreciated that they were worth preserving.[4]

The long history of plant images in manuscripts indicates their effectiveness in communication. Even some ancient Roman herbals were illustrated, and a copy of Dioscorides's text from about 512 CE has plants portrayed so naturalistically that they were copied in the late eighteenth century and used as a guide to Greek plants by a British botanist heading to the area. The problem with manuscripts is that there is no way to ensure that either text or illustrations will be accurately reproduced. As a result of repeated copying in the time before printing, there was a definite deterioration in the images in medieval herbals; they became so stylized that plants could not be identified from them.[5]

In an opposite trend beginning as early as the thirteenth century, naturalism started to creep into religious art; it became important to present God's creations as realistically as possible. This practice then extended to medical manuscripts. In the next century northern European artists painted more naturalistically; realistic plants appeared not so much in herbals as in oil paintings of religious subjects in flower-strewn gardens. Albrecht Dürer and some in the generation before him were masters of depicting plants. Art historians argue that this tradition led to the first good botanical illustrators, with Hans Weiditz, the artist of Otto Brunfels's herbal, perhaps having been trained by one of Dürer's students.[6]

The rise of naturalism in art seems to predate the careful, direct observation of nature that arose in the early sixteenth century in the medical community and may even have led to it, with artists and artisans encouraging closer attention to nature. The art-science relationship is often seen as one of support—art providing images useful in communicating science—yet there may be a more fundamental connection. Pamela Smith argues that early modern artists actually led the way toward seeing nature in a new light, which the botanists then exploited as they became less tethered to ancient texts. Difficulties in drawing can make for closer, keener observation, which early modern artists and naturalists saw as at the heart of inquiry. A quick glance was not sufficient. With the slow turning of sight into marks on paper came cognitive change, a different perception of forms and their relationship to one another. The practice of observation necessary for creating naturalistic art was then taken up by those asking questions about the natural world: artistic practice became essential to scientific inquiry.[7] Herbaria allowed botanists to look carefully at plants even when live ones weren't available.

A botanist needed time to get to know a plant, to have an extended conversation with it, and part of that conversation often entailed drawing. A relevant insight here was made by Agnes Arber, a plant morphologist and author of a book on early printed herbals. She contended that art is key to this area of science because there is much about a plant's visible characteristics that cannot readily be put into words: "Artistic expression offers a mode of translation of sense data into thought, without subjecting them to the narrowing influence of an inadequate verbal framework; the verb, to illustrate, retains, in this sense, something of its ancient meaning—to illuminate."[8] The benefit of drawing arises from examining details closely and then attempting to visually integrate them into the surrounding structures in the illustration. This requires repeated drawings to gain intimacy with the subject and also to train the eye and the hand. Arber knew this firsthand because she created illustrations for her research publications.

Because of the continued reliance on ancient authors, images were well ahead of texts in presenting accurate botanical information, making art key to the development of early modern botany, at a time before there was an adequate vocabulary to describe plant structures. As more and more plants were described, the distinctions between them became finer and so did the renderings. In exchanging drawings, botanists were often using them as surrogates for the plants themselves, as carriers of information. In fact, simply having images juxtaposed with the text eventually led to more correct descriptions in line with what the reader was seeing. In the early modern era, art became a tool of science, of learning about the world, and it remains so. It is not uncommon for a botanical illustrator to point out a structural feature that might have been missed by the botanist describing the species. In the recent case of a plant known only from an herbarium specimen, an illustrator interpreted a structure in a way that the botanist did not think possible. When the plant was found growing in the wild, the artist's rendering proved correct. This example is hardly unique. Art is not just a way to communicate a discovery, it can be essential to it.[9]

Printing and Artistic Conventions

Making art more available for study was a major move forward in early modern botany. Movable type and book printing originated in the mid-fifteenth century, and herbals, books on medicinal plants, were among those produced in this technology's early days. At first, the illustrations were merely copies of inferior images not drawn from life. But by the 1530s, good drawings became available. Books printed on paper transmitted botanists' findings to a larger community and often included illustrations taken from drawings done on paper and transferred to wooden blocks. Woodcuts were used for relief printing. A trained artist traced the drawing onto a block, then a wood cutter carved out the background, while the parts to be inked remained level with the surface, just as with the movable type, which was set on the same plate with the block. In herbals, the positioning of illustrations mimicked that in manuscripts, with image and text on the same page.[10]

Figure 3.1a. Wood engraving of septfoil (*Tormentilla*) in *Herbarum Vivae Eicones,* by Otto Brunfels, showing the lower stems drooping. (Missouri Botanical Garden, Peter H. Raven Library)

The relationship between botanical illustration and inquiry was being worked out simultaneously with new printing techniques, so botanists and artists collaborated in exploring a variety of approaches. Most of Hans Weiditz's drawings for Otto Brunfels's 1530 herbal were apparently done from life, though a few seem flat, suggesting their models were preserved specimens.[11] Lack of fresh material may have been the issue, as it would remain, with later artists becoming skilled at creating lifelike renderings from dried plants. Besides their realism, another reason for thinking most of the images were made from life is that the plants display blemishes. Weiditz did not draw idealized plants, but ones that he himself might have gathered; a septfoil plant has wilted leaves and flowers, particularly at its base (fig. 3.1a).

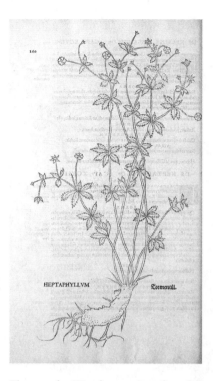

Figure 3.1b. Wood engraving of septfoil in *De Historia Stirpium,* by Leonhart Fuchs. (Smithsonian Libraries)

Twelve years later, the artist for Leonhart Fuchs's herbal took a different tack and depicted plants that were damage-free, "perfect" specimens, and sometimes ones that displayed fruits and flowers on the same plant even when these didn't occur at the same time in nature. The images in Fuchs were also simpler, making it easier for the eye to zero in on specific traits (fig. 3.1b). Not too many leaves were included, and the plants were somewhat flattened, almost like herbarium specimens, so the structures were spread out and didn't overlap too much.[12]

These artistic conventions are still used today. Botanical illustrations, particularly the pen-and-ink drawings employed in scientific publications, are realistic, but they are not hyper-realistic. The aim is intelligibility, to make the structures clear to the viewer, presenting a middle ground between naturalistic and stylized drawing. This relates to the history of herbaria, because these publications were produced in the same time frame that the herbarium became a scientific tool, a new way to observe plants. Like specimens, printed images offered a way to accurately communicate about plants. Thanks to printing, there was no longer the progressive deterioration in images that came with manual copying.

Visual Experiments

In *Eye for Detail,* Florike Egmond looks at several early modern collections of drawings of plants, animals, fossils, and minerals, what she terms "naturalia." Most were never published; they display novel drawing experiments that only later appeared in printed illustrations. Some are based upon new technologies, such as the magnifying glass, and new forms of inquiry, such as dissecting flowers and other structures. Egmond argues that by studying entire collections, patterns of presentation and thought become more obvious.[13]

In botanical drawings, plants were usually portrayed in isolation, decontextualized against blank backgrounds; they were rarely depicted surrounded by other plants in a natural habitat. This was not a new approach since medieval herbals presented plants this way, making them easier to identify. Isolating a single specimen is so ingrained a tradition in scientific illustration it seems hardly worth mentioning, but it has consequences. All seeing is theory-laden, and scientific illustrations can be considered composed scenes, leaving the impression that a plant can be understood in isolation, that its habitat is of little significance.[14] The white or cream background, the color of the paper, adds to this sense. It's not surprising that the same tradition is found with specimens; it seemed better to have the plant itself as the focus, though some botanists pasted several specimens on a page, not only for economy but for ease in transport: paper in bulk is heavy. With time, this doubling up became less common.

Egmond gives special attention to Conrad Gessner's work because his images provide examples of several art and research techniques. At times he drew arrows or lines to show that a notation relates to a particular part of a figure, thus linking text and image, or he displayed a sequence of images illustrating the development of a seedling, painting the stages close to each other so they would read as linked. His combination of drawing and writing helped to guide attention and control sight, indicating that a scientific argument, such as the idea of a species, is an interplay of the visual and verbal (see fig. 2.2). Gessner's drawings are full of close-ups of plant parts such as fruits, seeds, and flowers, or even floral structures. He seemed to be attempting to get to know a species through closer and closer examination and discussion. He sometimes included enlargements of structures, and there is evidence he used a magnifying glass. He was not alone in this. Drawings by the Italian botanist and artist Federico Cesi support comments in his manuscripts that he too employed a hand lens and, in examining pollen grains, a microscope.[15]

Gessner's notebooks were working documents revealing his study methods. The viewer can sense the passion in Gessner's studies, the thrill he must have felt as he added a new piece of information or drawing to a page. On many sheets there are pasted slips of paper with added observations, an information technology used by many naturalists. In contrast, most of the other drawings Egmond discusses are more finished works, such as those owned by one of Clusius's patrons and now bound in volumes called *Libri Picturati*.[16] The notations are sparse, consisting of little beyond the plant's name and perhaps a line about where it grew or its medicinal uses. The annotations in Aldrovandi's and Cesi's drawing collections are equally terse. What links the collections is the naturalism of the art, suggesting this approach was accepted by botanists and collectors as essential for documenting plants.

While some visual techniques Egmond describes eventually became part of the botanical art repertoire, others were less successful. Besides using herbarium specimens as models for botanical illustrations, and illustrations as supplements to the information provided by specimens, there were also several visual experiments in the early modern period in which the two techniques were combined. One method of limited use was nature printing—in its simplest form applying ink to a flattened plant and pressing it onto paper to leave an impression. It was a way to preserve detailed information from the plant itself without necessarily preserving the specimen: no bulky pages, no fear of insect damage. The Italian botanist Fabio Colonna, one of the first to use engraving on metal plates rather than woodcuts for illustrations, created a collection of nature prints. Some had details added in ink, and a number became models for the engravings in his book on basic botany.[17] Engravings had to be printed separately from the text, but they provided higher-quality, more refined images that became the norm in botanical publications for many years. Nature printing, on the other hand, was only used from time to time. Books of nature prints were later published, but this technique never became a significant scientific tool. Not as much information was preserved as with a specimen; flower parts in particular did not print well.

Another method was even less successful than nature printing. In a manuscript from 1560, there is a copy of an illustration from Fuchs's herbal of Solomon's seal, *Polygonatum,* that is colored and has three leaves of the species pasted to it, perhaps the maker's attempt at a more realistic image (fig. 3.2).[18] Adding the leaves gave a better sense of their texture and substance, of the plant's materiality. But this collage technique never became common, though to this day, people press leaves between the pages of flora or field guides with the entry for that species to augment the printed information.

In the Swiss botanist and physician Felix Platter's herbarium, a few of the specimens were "enhanced." For tulips, he pasted structures found inside the flower on the outside of the petals to make them visible. A bulky sunflower was taken apart, its elements glued individually to the sheet, and then the stem pasted over them, so the flower is seen from the back, with the frontal view in an accompanying illustration. For a *Campanula* specimen, Platter went a step further to present the flowers' color. He left one of the original flowers in place, with its rich blue color faded to a dull yellow. For the others, he took *Delphinium* blooms, which kept their color, cut them down a bit to resemble *Campanula,* and pasted them on, making it difficult to see the trick. The pains Platter took in creating these alterations suggest how important it was to him to present the plant as lifelike.[19] The duplicity of these techniques is what kept them from becoming common practice.

Figure 3.2. Illustration of Solomon's seal (*Polygonatum latifolium*) with leaves pasted on, copied from Leonhart Fuchs's *De Historia Stirpium*. (© The Trustees of the British Museum. All rights reserved)

They are used in decorative pressed flower arrangements, but not by serious botanists, who would cringe at such defilement of specimens.

Eight of the eighteen volumes in Platter's herbarium survive, their contents revealing what he valued in plant information. There are eight hundred specimens, 141 watercolors, and 650 wood engravings cut out of books and often colored. In many cases, Platter pasted a drawing, a print, or even both on the left-hand page with the specimen opposite, so he could examine different types of evidence at once. Among the drawings are cutouts of the original Hans Weiditz watercolor illustrations for Otto Brunfels's herbal, which had the first accurate botanical illustrations. Indicative of paper's value, Weiditz had painted on both sides of the sheets. Platter went to the trouble of cutting out the drawings to salvage parts of the images on each side. He considered them more as information than as art, willingly mutilating them in a way that would horrify art historians.[20] Platter's habit of juxtaposing illustration and specimen was used by a number of botanists to provide more information than either alone could.

A manuscript created by the German apothecary Johannes Harder in about 1595 represents yet another experiment. The bound herbarium has four hundred pages, with plants pasted on both sides of each sheet. Harder's father Hieronymus, a schoolmaster and botanist interested in medicinal plants, had created a number of these collections, many of which are still extant. The Harder herbaria are particularly interesting because missing portions on many specimens were painted in, but Johannes carried the technique further than his father, with additions to almost every page. Absent petals were added with an attempt to match the color of the living flower; bright pink or blue paint was in stark contrast to the brown of the dead flower and served as a visual notation of what the flower originally looked like. In other instances, missing leaves or portions of leaves were painted in.[21]

In some cases Harder added an entire flower. This practice solved a problem for those collecting specimens. Since the flower is often the plant part with the most information, it is considered important to include either a flower or its subsequent fruit when making a specimen. However, if a plant is collected out of season, then all that's acquired is usually a leafy branch. Harder's solution combined art and science; it was a hybrid, a collage of specimen and drawing. There's a tomato plant with a juicy ripe fruit painted on the specimen that was in flower. Also supplied was the base of the stem with roots beneath, an addition he made to many specimens. There are many cases of painted roots in the Harder herbarium. In later herbaria and botanical illustrations, roots were less commonly included. Roots were bulky and messy to deal with in mounting specimens. Since there was usually little variation among them, it didn't seem to make sense to add them to drawings. However, roots were also the source of many medicines, so they would have been important to include for Johannes Harder, an apothecary. Bulbs were another frequently painted addition because they were impossible to press; preservation of plant parts in alcohol or other media was well in the future. Harder was obviously experimenting with a genre here, a way of increasing the information in an herbarium by including art as notation.

In some cases, Harder incorporated ecological data—a fern growing against a brick wall, a fungus on an oak tree, a swamp plant mounted over ripples of blue water (fig. 3.3). The most common additions were in the habitat category, though such were hardly informative. Harder would portray a generic grassy ground or tuft from which the stem arose, seemingly as a reminder of the plant in its natural state. The grass hints at the plant's place in a medicinal garden. The younger Harder's herbarium was probably used in communicating with apothecary customers to help identify useful plants and to display ones new to the area, such as the tomato and tobacco. Grounding the plants could make them appear more familiar and attractive to his audience.

Dried plants may not be the most eye-catching items as they fade to an almost uniform brown color, so Harder added a splash of bright paint to enliven the specimen or a decorative detail such as a tuft of grass. His was not a printed catalogue for distribution but one that customers were shown when visiting his shop, or that he could carry to them. This points to using herbaria for commercial purposes. A little later, as the nursery trade developed, herbaria served to show potential buyers the "real" plant they were investing in.[22] As new uses arose, new styles developed, making herbaria more than just informative—they were created to be pleasing to the eye as well. Nursery catalogues were printed, often on good paper, so they could be saved for reference.

Figure 3.3. Johannes Harder's specimen of marsh mallow (*Potentilla palustris*) with painted additions of a flower and blue water ripples as clues to the plant's habitat. (Oak Spring Garden Foundation, Upperville, VA)

Over time, herbaria were also used as presentation pieces, as some of the elder Harder's were. These were for individuals more interested in learning plant names than studying them closely and for those with the money and power to support the work of the "real" botanists. Aldrovandi's herbarium is spare, plain sheets of paper with the mounted plant and its name, as would be expected of Luca Ghini's protégé, and Harder included what he saw as needed information: the name in Latin and German. Soon the addition of stylistic flourishes like page borders, which came out of the manuscript tradition, began to appear and became more elaborate. Sometimes the border was not drawn in; rather, a paper trim was added that also reinforced the sheet (see fig. 4.1b).

Other decorations were more fanciful: vases pasted over the cut end of the stem, ribbons attaching the plants to paper, and names written in elaborate calligraphy (fig. 3.4). However, painting in missing parts of plants or adding background never became an important part of herbarium practice, though it did occur from time to time. In the eighteenth century, a French botanist drew in grounding for his moss specimens, and later the polymath Abbé René Just Haüy matched leaf and flower colors with pieces of colored paper that he cut out and used to enhance faded specimens, similar to the way in which the Harders used paint.[23]

Art and science have had a long relationship in herbaria. Botanical illustrations, some pasted on sheets, were frequently stored in herbaria along with the plants they depicted so they could be consulted together, as with Blanche Ames's orchid watercolors. Drawings were sometimes placed in separate folders to protect them from damage from plant chemicals but leaving them still readily available to supplement the information in the specimens themselves. This practice became less common, and most illustrations are now stored more safely in botanical libraries or archives. However, the newer practice implies a separation of art and science, a signal that perhaps the images are more about aesthetics than documentation of observation. This perpetuates an undercurrent of devaluation of the visual that has long influenced the thought of some botanists, creating a tension among practitioners regarding how best to communicate about plants. But there is

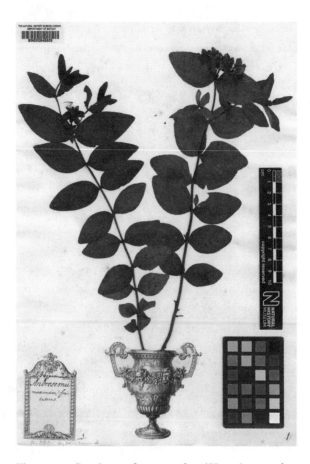

Figure 3.4. Specimen of sweet-amber (*Hypericum andro-saemum*) with elaborate label and vase, from the George Clifford Herbarium. (Museum of Natural History, London)

no "best" way; as Richard Mabey writes: "The quintessence of a plant can only ever be a fantastic goal, something to travel towards but never reach."[24] Refinements of techniques continue to this day, and each age brings its own values to the enterprise. The next chapter deals with how information about new species in the form of specimen/text/image was acquired by explorers and communicated to Europeans eager to learn about them for a variety of reasons: scientific, economic, political, and aesthetic.

FOUR

Early Exploration

AS CAROLUS CLUSIUS'S WORK INDICATES, European bota-
nists thrived on describing new plants, documenting what we now think of
as biodiversity. In the sixteenth century this could often mean delving into
areas relatively close to home: mountainous regions, deep forests, wetlands.
The work continued and became more organized in the next century, particularly regarding the hunt for useful plants. One German prince had his
foresters survey for medicinal plants and collect specimens, seeds, and information on them.[1] But as such plants became more familiar, as specimens
and seeds spread through botanical networks, botanists craved more novel-
ties and the search moved farther afield. This quest coincided with voyages
sponsored by monarchs seeking new sources of wealth and power.

Sixteenth- and seventeenth-century European explorers were overwhelmed by what they met in their travels and could only scratch the surface of what they saw. The species were so unusual in this "new" world that
European botanists were perplexed by the flora, to say nothing of attempting to decipher a plant's uses. Many explorers had scant scientific knowledge and were looking at totally unfamiliar species. They had large areas
to investigate and sometimes could learn little from indigenous peoples due
to language barriers and varying degrees of mutual distrust. Yet the search
for new plants and animals was intense, driven by a thirst for novelty and,
more important, a desire to obtain medically and economically worthwhile
species.

Middle Eastern Flora

The Middle East had long been accessible to those trading in spices, silk, and other luxuries. But in the early modern era, the prime reason for the botanically curious to travel east was to visit regions studied by the ancient writers Theophrastus and Dioscorides. Travel to the source of the plants they described from the eastern Mediterranean region became more pressing in order to compare them with western European species, which in some cases were considerably different. In addition, though spices had been traded since ancient times, by the mid-sixteenth century the Portuguese, with an outpost in India, were exacting high prices for Indian pepper and cinnamon grown in Ceylon, as well as cloves, nutmeg, and mace from the Molucca Islands in Southeast Asia. This made the search for alternate sources and trading sites attractive.

In 1547, the French king Francis I sent the naturalist Pierre Belon on a diplomatic mission to the court of the Ottoman Empire in Constantinople. Before reaching Turkey, Belon collected specimens and seeds in Greece and Crete for several months, then went on to Egypt, where he described doum palms, sycamore figs, and cassia trees with bark that yielded a form of cinnamon. Although Belon's specimens were lost to pirates—one of the many dangers facing explorers—he managed to return to France with seeds, which he coaxed to grow. Seeds were a good way to transport plants. They were more likely to survive than seedlings or cuttings, and if they germinated and flourished, then specimens could be made and more seeds distributed to others. Before the age of heated greenhouses, plants native to warmer and drier climates were difficult to grow, but Belon was one of the earliest to successfully acclimatize some exotic species.[2]

Financed by a merchant interested in developing trade in medicinal plants, the German botanist Leonhard Rauwolf set off in 1573 for the Middle East with a fellow physician. Rauwolf wrote of the cold and heat of the desert, the difficulties of a long trek on a camel, the threat of robbers, and encounters with local warring factions. He found many plants described by the ancients growing in their native ranges. A volume of specimens Rauwolf collected in the Middle East still exists. Among the plants are cultivated varieties of chickpea, sorghum, cotton, and eggplant, all grown over 440 years ago and long supplanted by other cultivars. When an Israeli researcher recently studied the specimens, this was the first time someone from the region where they were gathered had set eyes on them since they were collected. They are invaluable historical documents and may yet yield genetic data. In all, there are 191 species in this herbarium volume, of the 364 plants Rauwolf described in his journal. The book he wrote on his return contains illustrations of 42 of them, some apparently based on the specimens (figs. 4.1a, b).[3] The herbarium pages are edged with decorative paper to strengthen each sheet and give it distinction, a practice that appears from time to time in old collections.

More than a century later, the French botanist Joseph Pitton de Tournefort traveled east early in his career, accompanied by the artist Claude Aubriet to draw the plants he collected. This practice later became more common among explorers; it was important to return with seeds and specimens, but images of living plants provided the best record of plant form and color. There had been so many cases of travelers losing their collections that Tournefort took the precaution of keeping duplicates of all his finds so each set could be sent back separately.[4] Tournefort successfully brought back over 1,350 plants, still preserved in the herbarium of the National Museum of Natural History in Paris.

Latin America

The explorers who headed across the seas have received more attention than those who went to the Middle East since the former moved into territory uncharted by Europeans and pushed the boundaries of navigation and seamanship. The Spaniard Gonzalo Fernández de Oviedo y Valdés published the first account of Western Hemisphere natural history in 1526. A second, larger version was illustrated with thirty-five woodcuts.[5] As with most natural histories on early European explorations, it dealt not only with plants but with animals, minerals, geography, and the cultural uses of natural materials, especially food plants.

Oviedo was among those fascinated by the cacao tree, the source of chocolate, and he also described the avocado, banana, and papaya. He

Figure 4.1a. Specimen of eggplant (*Solanum melongena*) with flowers and fruits, collected by Leonhard Rauwolf in Syria. (Naturalis Biodiversity Center, Leiden)

Figure 4.1b. Illustration of eggplant, also with flowers and fruits, in Rauwolf's published travel journal. (Getty Research Institute, Los Angeles)

introduced the pineapple to Europeans, discussing its delicious sweetness as well as its odd structure.[6] Since the earliest known herbarium dates from about 1540, it's not surprising that Oviedo did not collect specimens, and in any case a pineapple fruit would have been impossible to press between sheets of paper. With botanical terminology still in its infancy, it was difficult to adequately describe unusual plants. Though some seeds, seedlings, and cuttings reached Europe from Columbus's time on, and corn, squash, and papaya were soon growing in Europe, other species, such as pineapple and banana, required specific watering regimens, warm temperatures, and soil types difficult to replicate.

King Philip II of Spain sent Francisco Hernández de Toledo, his personal physician, to the New World in search of medicinal plants. Hernández landed in Mexico in 1571 and toured the country as well as other areas of Central America until 1577. He worked with a team that included a geographer, artists, botanists, and over twenty indigenous medical practitioners. Hernández was more than willing to seek native expertise and to credit it. Though local knowledge, to say nothing of local labor, was not often acknowledged by explorers, it was always drawn upon because newcomers knew little of an area's geography, biology, or culture. He noted the name of each species in the local Nahuatl language and in Latin, gave its use, and had indigenous artists document the plants.[7]

Hernández amassed four thousand illustrations and twenty volumes of notes along with seeds and an extensive herbarium. When he returned to Spain with this material, the king was thrilled, but thought that Hernández was not up to making order out of it since the collection was in disarray. The monarch gave the task to his new physician, the Italian Nardo Antonio Recchi, triggering a complex and lengthy process that ultimately led to the publication of only a portion of Hernández's hoard. Recchi did not consider Nahuatl names and indigenous information useful to Europeans, often the fate of such data. In addition, he chose to focus on species that appeared similar to European plants, though in many cases they were not closely related, a practice repeated many times by botanists attempting to make sense of strange plants. The unfamiliar species were often termed *herbae nudae*, or bare plants, ones not described in ancient texts. Later research suggests that Recchi included descriptions of just six hundred of Hernández's three thousand plants.[8] This approach, not surprisingly, caused conflict with Hernández. Recchi returned to Italy in 1583, bringing with him a copy of his manuscript and six hundred illustrations produced from Hernández's drawings, but he never published the work.

After Recchi died in 1594, the manuscript passed to his nephew, who had no means to publish it. Finally, a group of Italian naturalists set their sights on obtaining the Hernández/Recchi manuscript, appreciating the novelty and importance of this strange flora. They finally finished publishing it in 1651, eighty years after Hernández's arrival in Mexico.[9] Each entry in the book began with the Nahuatl name of the species, making it the largest botanical glossary of non-European names available at the time. Though the volume represented a limited portion of Hernández's original material, it became particularly valuable after most of his documents, including his herbarium, were destroyed by fire in Spain's El Escorial in 1671, which explains why there are no extant specimens linked to Hernández's writings. Sagas like this, of delayed publication, dispersal of documentation, and disaster, are far from rare in the history of botanical collections.

Another approach to studying American plants was taken by Nicolás Monardes, who never set foot there. Instead, he studied these species by growing them, observing them closely, making specimens from them, collecting seeds, and distributing these to correspondents throughout Europe. A physician, he also experimented with preparing medicines from the plants he grew, with his observations going into his written descriptions. While writers like Oviedo discussed natural history broadly, Monardes focused on plants, particularly medicinal ones. Though his three-volume work sold well in Spain, translations gave it wider appeal. Carolus Clusius published a Latin edition, making it available to literate classes across Europe, and John Frampton produced an English version with the wonderful title *Joyfull Newes out of the Newfound Worlde*.[10] All editions were illustrated, another reason for their popularity.

Though the Spanish and Portuguese dominated the first century of Latin American exploration and colonization, other nations soon became interested in the riches found there. The Netherlands became a naval power and in 1624 wrested control of northeast Brazil from the Portuguese. They then financed a study of the area's natural history, a more organized scientific and governmental venture than any seen earlier. The expedition included a physician, a naturalist, and two artists. In 1648 *Historia Naturalis Brasiliae* became the first major illustrated natural history of any region of the Americas whose text and illustrations drew from firsthand experience.[11] Observations included Portuguese, Spanish, and indigenous names for each species. Including indigenous vocabulary was a way to suggest the research's depth and to link these plants to their local uses.

The expedition's plant specimens are among the earliest from the tropical New World, though an earlier collection, created around 1587 by a Dutch collector working further west in South America, in what is now Suriname, preserves plants native to the area and also the African food plants okra and sesame. This is evidence that the plantation culture, with the presence of African enslaved persons, had brought with it new species, one of many examples of the early movement of plants with links to the slave trade. It suggests how herbaria can contribute crucial evidence regarding cultural and political history and can help clarify portions of this history that have long remained hidden, including the early pervasiveness of enslaved labor in the Americas.[12]

North America

North American voyagers faced different hardships than those in the tropics: cold and snow rather than heat and humidity. In the early days of European exploration, the Portuguese and Spanish had a head start, especially in the warmer areas of the New World, so the French and British aimed at North America. Exploring what is now Canada, Jacques Cartier sailed to Newfoundland and then on to the gulf of the St. Lawrence River on the first major French expedition there in 1534. On his second trip, he traveled down the St. Lawrence River to what is now Montreal, where his ships were stuck in ice for the winter. The French survived the ordeal in part because First Nations people told them of the restorative powers of a cypress Cartier called arborvitae, tree of life, since a tea made from it cured his crew's

scurvy. After this harrowing experience, Cartier returned to France, bringing seeds of several species from the over thirty he described in his report.
He traveled at a time before herbaria had become common, but his seeds,
including arborvitae and snakeroot, were sown in Paris and distributed to
others in Europe along with specimens.[13]

In 1635 French botanist Jacques-Phillippe Cornut described about
sixty Canadian plants. These descriptions were based on plants grown
from Canadian seeds by Jean Robin, the king's gardener, who distributed
the seeds to many botanists. Robin and his son Vespasien traded seeds with
Carolus Clusius, the great collaborator, botanist James Garret and nurseryman John Tradescant in England, Cardinals Farnese and Barberini in
Rome, and Caspar Bauhin in Switzerland, who described several of these
plants.[14]

By the mid-seventeenth century, there were several northern settle-
ments. At this time, the nurseryman John Tradescant Jr. made three collecting trips to Virginia. He complained of difficulty identifying plants without
reference books and of the language barrier making it hard to communicate
with Native Americans. However, he returned with about two hundred species, and among those introduced into British gardens were the sycamore,
tulip tree, and black walnut.[15]

A notable Virginia collector was John Banister, who had studied at
Oxford and was the first university-trained botanist to send specimens back
to England. He was encouraged by Bishop Henry Compton, an avid amateur botanist and gardener, who selected him for the clerical post based on
his botanical knowledge. Before leaving for his assignment, Banister studied specimens already received from North America, including many grown
from the seeds collected by French explorers. He brought with him an herbarium of such plants and left a catalogue of them in Oxford so that botanists there would know what specimens he was referring to, thus taking a
more systematic approach than other collectors of the time.[16] Banister's
specimens were well received by British collectors and botanists, who
rushed to describe the new species. So extensive was their use of his material that he vowed to write his own book, but his life was cut short by an
accident, one peril of plant collecting in the wilderness.

India

As with the Americas, the lure of the East drew many nations in search of wealth. Today it's difficult to appreciate how valuable spices like nutmeg, cinnamon, and cloves were from ancient times. The plants they came from grew only in areas of the Middle and Far East, but because spices were dry and could effectively be used in small quantities, they became commodities worth trading over long distances. As commercial enterprises grew in the early modern era and navigational innovations developed, longer sea voyages were undertaken. The Portuguese reached India's west coast at Malabar, home of pepper, and managed to monopolize the market for this spice in the early sixteenth century. They also introduced a number of South American species not only into Europe but into their Indian colonies, and soon such plant exchanges became widespread.

By the seventeenth century, Portugal's dominance in the spice trade had diminished, as the Dutch gained control of the Malabar Coast. Other Europeans were also settling in the region. Since the vegetation was unfamiliar, the only recourse, beyond trial and error, was to learn from indigenous peoples. There were major problems of distrust, made worse by lack of a common language, but those who remained for some time and were willing to be patient often learned much from indigenous peoples, especially those adept at medicine. While traders and the physicians who accompanied them might not be in the area for a long period, missionaries often were. Treating illness was a way to connect with natives, and for some missionaries, studying plants was an interesting leisure activity.

India was seen as ripe for evangelization by the Catholic Church as it sought to extend its sway after the Protestant Reformation in the sixteenth century. The largely Catholic countries of Spain, Portugal, and France supported these efforts in opposition to the largely Protestant British and Dutch. While Catholics and Protestants might not treat each other warmly in Europe, in India their shared culture outweighed their theological differences— perhaps especially when they had a mutual interest in plants. This was the case with Hendrik Adriaan van Rheede, the Dutch governor of Malabar, and the Italian priest and physician Matteo di San Giuseppe. Because of his government position, van Rheede traveled throughout Malabar, and though he was not a botanist, he collected plant specimens and information on their uses.

San Giuseppe had been in the area for years and had amassed a large specimen collection with notes on indigenous plant names and uses. He was attempting to deal with a serious local need for medical supplies. Shipments from Europe were sporadic at best and had often lost their potency or were ineffective against the different ailments common in what Europeans considered an alien environment.[17] After years of study, San Giuseppe was happy to share his plant knowledge and materials with van Rheede, who planned an illustrated guide to Malabar plants. The priest even had drawings of many species.

Van Rheede investigated San Giuseppe's cache and found it not in publishable form. He assembled a team of native artists as well as translators, collectors, and a chemist to produce better content and set up a board of European and indigenous physicians. They provided plant names in Malabarese, Sanskrit, Arabic, and Latin, and agreed on the plants' medicinal uses. This massive project ultimately resulted in the twelve-volume illustrated *Hortus Indicus Malabaricus*.[18] The art presents the plants clearly, but not always in enough detail to identify their species today, especially since van Rheede's herbarium has not survived for comparison. The style is definitely a hybrid of Western botanical illustration and Indian art (fig. 4.2). Such a blend is seen frequently when different cultures collaborate on picturing plants, revealing the subtleties of depiction.

Not only did van Rheede have to work with a large group in India, he also had to negotiate long distance, working with botanists and printers in the Netherlands. In botanical exploration, there was often a fraught interplay between professional botanists at home, who considered themselves the taxonomic experts, and observers in the field, like van Rheede, who believed botanists were skewing the descriptions to emphasize plants' relationships to European species. Experts had different perspectives because they had access not only to the specimens sent back but also to their own collections. There were divergent views in part because they were looking at different evidence.

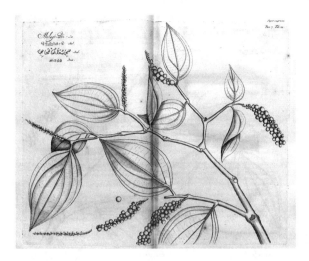

Figure 4.2. Illustration of black pepper (*Piper nigrum*) in Hendrik Adriaan van Rheede's *Hortus Indicus Malabaricus*. Though the illustrations were produced in the Netherlands, they were based on drawings sent from India. In the upper left-hand corner the plant is named in four languages. (Missouri Botanical Garden, Peter H. Raven Library)

When the British arrived in India, their physicians also sought useful medicines. In Madras, they worked with Tamil doctors and apothecaries, again feeling their way toward closer collaborations over a period of time. British doctors created gardens so they could study the plants over time and experiment with different kinds of preparations, sometimes selling these to patients and to other physicians. These men also supplied specimens to botanists in Britain eager to describe new plants: publishing on new species became a form of status for those interested in botany. This is how a few herbaria in cities like London and Edinburgh as well as Paris and Leiden became "centers of calculation," focal points for collections and for creation of knowledge about the widespread regions being explored. Recent observers have critiqued this concept, citing work like van Rheede's. Those in India generated and used a great deal of natural history data, sharing it with Europeans, but it was hardly a one-way street with all the expertise shuttled back home.[19]

Eastern Asia

Repeatedly, the history of herbaria moves into stories of commerce and political control. As they had in northern Brazil, the Dutch followed the Portuguese into the East. By the beginning of the seventeenth century, the
Netherlands was a major commercial force, operating through the Dutch
East India Company, which set up trading outposts not only in India but in
what is now Sri Lanka, Indonesia, Malaysia, China, and Japan. This gave
the country control over trade with remote areas like the Molucca Islands,
at that time the only source of nutmeg, mace, and cloves. The company
employed physicians to care for their far-flung teams at risk of injury, poor
nutrition, and infections. One was Paul Hermann, recruited as a medical
officer for Sri Lanka. On the way, his ship stopped at the Dutch outpost of
Cape Town, South Africa, and he became the earliest known collector
there. He sent specimens and seeds of twenty-two plants as well as the first
list of African plants back to Europe to protect against their loss before he
traveled on.[20] Plant lists became important historically because they suc-
cinctly document an area's flora, in Hermann's case providing the first record of sub-Saharan Africa's biodiversity.

Hermann spent five years in Sri Lanka studying its flora, amassing
hundreds of specimens and illustrations, and investigating the cultivation of
true cinnamon, *Cinnamomum zeylanicum,* which was native to the island
and parts of southern India. A Chinese species, *Cinnamomum cassia,* has
slightly different characteristics, and working out the distinctions took
years, as botanists had to rely on specimens and often unclear descriptions.
By the time seeds or specimens of new plants reached a European expert,
their labels and descriptions may have been lost or switched, and even their
origins became uncertain; a few South American species were first described as coming from Africa. This problem persisted for centuries, frequently complicated by economic interests intent on finding the "true"
source of a species.[21]

There are four volumes of Hermann's specimens and one of illustrations at the Natural History Museum, London (NHM). All the hands
through which they passed on their way there are not known. In fact, after

Hermann's death in 1695, they did not surface again until 1744, when they were owned by the Danish apothecary-royal, who passed them on to the Danish lord chamberlain.[22] His heirs sold the collection to a professor, who then sold it to the British botanist who donated it to the NHM. Journeys like this were not uncommon for old herbaria, nor was selling them, indicating how valuable they were considered, especially a collection like Hermann's that contained so many exotic plants.

Also in the employ of the Dutch East India Company was Engelbert Kaempfer, a German-born physician and plant collector who traveled in Persia and then worked in Nagasaki, Japan. Aside from this port, which was open to Dutch and Chinese ships, Japan was essentially closed to foreigners. Traders were confined to the port—to a man-made island built for the purpose. Kaempfer, undeterred by restrictions, even hunted for seed in fodder shipped in for animals. Once a year the Dutch traveled to the capital at Edo for an audience with the shogun. Kaempfer made the most of this opportunity, botanizing along the way and learning about Japanese life in what might be considered an early form of fieldwork. Plant collecting involves more than just pressing plants; curiosity and social skills are important in picking up local knowledge.[23]

Kaempfer was the first Westerner to describe the ginkgo tree and send specimens and seeds to Europe; he also wrote of camellias and rhododendrons growing wild in mountainous areas. His *History of Japan* includes a section on plants central to Japanese culture. Tea is one obvious example, but he also described mulberry leaves for feeding silkworms and the importance of giant radishes in the Japanese diet. Kaempfer remained in Japan for two years, returning to Amsterdam in 1695. He published a work presenting Japanese natural history as well as material on his stay in Persia. He drew not only on what he had learned about Japanese plants firsthand but also from important Japanese natural history texts, such as its first illustrated encyclopedia. Many of his illustrations were based on this work, another example of East-West artistic interactions.[24]

When Kaempfer died four years later, his notes and specimens were sold to Hans Sloane, the great British collector. Sloane's herbarium is now at the NHM and is one of the largest collections of plants from this period, consisting of 265 bound volumes. Volume 211 contains Kaempfer's plants and is among the most important in the herbarium because Sloane also bought the accompanying Kaempfer manuscripts, including original drawings with Japanese plant names and references to botanical works. Tying herbarium sheets to information increases the specimens' value because it gives them context.[25]

One last Asian explorer deserves mention. William Dampier was a British naturalist who traveled for commercial purposes, operating on the fringes of legality. He has been called a pirate, but technically he was a privateer. His activities, although sanctioned by Britain, were considered acts of theft by other nations whose ships or colonies he raided. On these high seas' adventures he explored the natural history of areas he visited and even collected specimens. Dampier was the first to return from Australia with pressed plants, gathered on his second trip there in 1699 and hinting at the unique flora of the continent.[26] His ship sank, but he was able to save the specimens, an indication of how seriously he took natural history. A crew member had painted a number of the plants and animals Dampier collected, and these along with the specimens are still extant, a testament to the lure of the natural world even for a swashbuckler.

Travails of the Collector

Dampier's saga points to the hardships endured by plant collectors, particularly in the early days, when shipwrecks and other disasters were commonplace. However, traveling to remote areas and dealing with political, physical, and "Kaempfer's"? cultural difficulties remains challenging. Since the greatest number of species—the richest biodiversity—is in the tropics, these areas have long been magnets for collectors. But biodiversity means not only many plant species but poisonous animals, parasites, and disease-carrying insects as well. Insects are ubiquitous and harbor infectious organisms that cause yellow fever, malaria, and scores of other problems, making them the most bothersome and dangerous animals encountered, though menacing snakes, tigers, and crocodiles add spice to exploration. Today, botanists are armed with an extensive pharmacopoeia and fortified with vaccinations and pest repellants. Still, they suffer from infections as well as assaults by biting insects, mites, and ticks, to say nothing of leeches and snakes.

Insects can also devour specimens. In addition, humidity in much of the tropics makes specimens difficult to dry. When they could not be desiccated or kept dry, entire collections were lost to mold and decay before being shipped home. The Austrian botanist Nikolaus Joseph Jacquin turned to drawing rather than preserving plants when his Caribbean collections rotted; others resorted to nature printing for the same reason (see fig. 7.2).[27] Whether for drawing, printing, or pressing plants, paper was a necessity and often one of the most precious commodities on expeditions as well as one of the heaviest to cart around. But even when the specimens were successfully packed for shipping, dangers lurked, with shipments lost at sea to piracy, shipwreck, or water damage. Even if they had survived disease, accident, and encounters with unfriendly humans, many collectors did not fare much better on the trip home.

Even today, it takes stamina and nerve to go collecting. There are mountains to scale, rivers to ford, thick brush to cut through. Climbing trees is frequently necessary because so many tropical plants are epiphytes living on other plants, including tall rainforest trees with enticing flowers and fruits dangling well out of a collector's reach. Before the creation of elaborate rope, net, and pulley systems that allow researchers to go high into the tree canopy and remain there for extended periods to study the flora and fauna, there were several approaches to this problem. While collecting in South Africa in the eighteenth century, Carl Thunberg shot down fruits from tall trees, and in 1930s Malaya, Edred Corner trained monkeys to collect from trees.[28] These extreme measures give some hint of the passion driving collection.

Many plant hunters thought the difficulties were worth it, some for the thrill of finding beautiful plants, others for the status or financial gain it would bring them back home. In Europe, plant specimens were keenly awaited by entrepreneurs looking for business opportunities, including new sources of timber, medicines, and garden plants. Wealthy gardeners were keen to have the latest discoveries and did their utmost to acquire and nourish them, while botanists were eager to describe new species and add them to their herbaria.

The Value of Collecting

WHILE THERE WERE MANY DRIVERS FOR exploration, curiosity was a major one for those amassing collections of plants, animals, minerals, art objects, and anthropological artifacts. Europeans had a thirst for the new and different, a desire to be among the first to see a tulip or a tobacco plant. It is no wonder that collecting became an obsession with those who combined a passion for knowledge with the means for acquisition. These were closely related since education was usually limited to those with wealth or with wealthy patrons.

Serious scholars like Ulisse Aldrovandi relied on collections to learn about the world by direct observation. In the case of plants, this meant studying plant structures, comparing species, and investigating medicinal properties. The more varieties available, the richer learning could be, and this was documented in Aldrovandi's extensive notebooks and collections. Status was another impetus for his acquisitioning. He opened his collection to others, both to learn from them and to display his erudition and resources. As often happens with passionate collectors, his assemblage grew so large that it overwhelmed his project. This was despite his hiring assistants, essentially curators, to make sure items were labeled as to what they were and catalogued as to where they were kept. He faced difficulties systematizing his observations to make sense of them and circulating his results, only some of which were published, most posthumously.[1]

Ordering items and observations can be laborious. A plant specimen must at least have its name attached, if not information about where and when it was found. The name would be given in Latin and one or more vernacular languages, sometimes with a reference to the species in the botanical literature. Then the specimen had to be stored in some order that made it readily available in the future. When specimens accumulated in the thousands the problem could become unwieldy. Many collectors lost control of organization, especially later in life, leaving difficulties for those who inherited or purchased their collections.

The Sloane Collection

Aldrovandi's museum was orderly enough that at least a portion of it has survived in good condition, the plants more so than animal materials, which are more difficult to preserve. An even larger collection was that of Hans Sloane, whose name has come up in relation to Paul Hermann and Engelbert Kaempfer's specimens from Asia. It is thanks to Sloane and his network that many specimens from seventeenth- and eighteenth-century plant collectors still exist. But his collection went far beyond flora or even fauna to include books, art, medals, and exotic artifacts. After his death, it formed the basis for the British Museum; this is how his herbarium came to be so carefully preserved. It is inevitable that any history is skewed by what evidence survives, by design or by chance. Sloane plays an outsized role in botanical history because he collected so many specimens and ensured their survival. However, because his collection includes plants from all over the world, it is somewhat representative of what was going on in botany at the time.[2]

Sloane held an esteemed position in British society as a physician to members of the upper classes, an entrepreneur who made chocolate in milk fashionable, and a custodian of an impressive collection that he displayed in his London home. Among Sloane's many spheres of influence was the Royal Society of London; he became a member in 1685 and eventually its president. The society was founded in 1660 just as a period of political and religious upheaval in Britain wound down. Members were dedicated to the ideals of empirical inquiry and experimental science proposed by Francis Bacon, whose advice was to collect sufficient information on a topic before drawing any conclusions about it—in other words, inductive reasoning. With this perspective ushering in a new era, it was a hopeful time.[3] Sloane exemplified that spirit, seeing collecting as information gathering. As had Aldrovandi, he aimed to raise his status as someone with the knowledge to build notable collections in several areas.

Sloane was also involved with the Worshipful Society of Apothecaries. When he moved to Chelsea, his property included the land on which the society's Chelsea Physic Garden had been created in 1673. The organization grew medicinal plants and had a laboratory for studies in preparing medicines. Eventually Sloane deeded the land to the society with the stipulation that each year it would send fifty medicinal plant specimens to the Royal Society for its collection. This was to ensure that the garden continued to grow such species, since there was an opposing tendency by some to put more emphasis on garden plants. The agreement was to remain in effect until the society received two thousand specimens, though the final total was thirty-five hundred, all of which still exist today.[4]

Relatively early in his career, Sloane served as physician to the British governor of Jamaica and took the opportunity to collect island plants and animals, relying on British expatriates, indigenous peoples, and enslaved persons for information. However, these sources are not always cited in Sloane's two-volume *Natural History of Jamaica,* which contains fascinating descriptions and illustrations of the island's flora and fauna. While in Jamaica, Sloane had a botanical artist paint watercolors of many living plants. On returning to England, he commissioned Everhard Kick to create the illustrations for the book. Kick worked from the watercolors and the specimens, skillfully blending information from both. There are still pressure marks on some specimens, evidence of Kick's tracing them. Reminiscent of Felix Platter's volumes from a century earlier, the drawings were pasted across from the relevant specimens, though in the case shown in figure 5.1 they were squeezed onto the same sheet. The arrangement was done by Sloane's curator in part to ensure that the two collections remained together after Sloane's death.[5]

Eventually Sloane's plant holdings grew so large he hired curators to organize them and make them available to botanists. He has been called a

Figure 5.1. Everhard Kick watercolor and Sloane Herbarium specimen of Florida Keys thoroughwort (*Eupatorium villosum*, now *Koanophyllon villosum*). (Sloane Herbarium, Natural History Museum, London)

"collector of collectors" with good reason; there are specimens from over 280 individuals in his herbarium. He traded specimens with many, sponsored overseas collectors like John Banister, and bought collections from Kaempfer and others. He also had ties with an avid circle at Oxford University, where John Banister, before going to Virginia, had studied under Robert Morison, who also had a large network. Jacob Bobart the Younger headed the Oxford Botanic Garden and taught at the university. He tutored Sloane in the importance of collecting specimens. Bobart himself traded information and specimens with botanists at the King's Garden in Paris and the university garden in Montpellier, long a center of botanical research.[6]

William Sherard worked with Morison at Oxford, collected widely, and knew Sloane and his London colleagues. However, the two eventually quarreled because Sherard would not lend some specimens to Sloane, who then refused to share his. Both Sherard and Bobart left their collections to the university herbarium. Donation to an institution was uncommon at the time, though it eventually became the lifeblood of many herbaria. Unfortunately, a later Oxford curator removed a number of collections from their bindings and at times labels were replaced, so it is impossible to know what arrangements earlier botanists had used for specimens: a case of what those in the herbarium world now call "curation crimes," losing valuable data in attempting to improve organization.

In London, Sloane socialized with a large circle of plant collectors, most notably at London's Temple Coffee House, where a group of the botanically minded met regularly. There Sloane was likely to run into men like Bishop Henry Compton, Banister's mentor, as well as Leonard Plukenet and James Petiver, both avid plant collectors who described many new species. Though the group shared much and collecting was definitely a communal activity, which was part of its attraction, rivalries at times cooled relationships and stifled information sharing.[7]

Petiver was the most driven of these collectors. As an apothecary, he was not considered to be in the same class as Sloane and Plukenet, who were physicians and moved in higher social circles, yet they all collaborated when it served their botanical purposes. Perhaps Petiver's passion, if not obsession, with collecting was his way to compensate for and attempt to elevate his inferior status. He was not interested merely in exotic plants; those from Britain and the continent were welcomed, and he participated in the Society of Apothecary's local field trips. However, he also worked hard at finding collectors to keep him supplied with newly discovered plants worldwide, and encouraged anyone he knew who was traveling abroad to send specimens: friends, customers, apothecaries, physicians, surgeons, merchants, missionaries, and so on.[8]

Like many avid collectors, Petiver wrote an instruction sheet on collecting to improve the quality of specimens. It included: "As amongst Foreign Plants, the most common Grass, Rush, Moss, Fern, Thistle, Thorn, or vilest Weed you can find, will meet with Acceptance, as well as a Scarcer Plant." He noted that plants in fruit or flower were more desirable, and that fleshy fruits should be sent in spirits or brine. Petiver was happy to provide his collectors with paper, jars, and other supplies, and was willing to pay for specimens. In some cases, he sent medicines for collectors' physical complaints. He also scolded them if they did not come through, and one collector was so angered by this that he sent nothing more. Materials often were lost in transit, and it was particularly frustrating to Petiver when letters arrived but specimens did not, the letters promising wonders he never saw. He acquired about one hundred herbaria and tried to organize this disparate collection along geographic lines, an early attempt at dealing with the relationship between plant traits and environment.[9] He frequently replaced labels, often with data loss. He was so overwhelmed by the number of specimens that when Sloane acquired the lot after Petiver's death, he was appalled by its condition.

Important sources for Petiver were the captains and surgeons on slave ships sailing a triangular route: from Britain to Africa, carrying goods that were sold there; then conveying enslaved Africans to be sold in Latin America and the West Indies; and from there transporting slaves to the American colonies. The ships then carried sugar, coffee, and tobacco from the Americas to Britain. Petiver's worldwide collection was shaped by the geography of slavery, with many specimens coming from West Africa, Latin America, the Caribbean, and the North American colonies. Slavery was intertwined with plant collecting over a long period. Plant collectors often engaged with enslaved and indigenous peoples to obtain information on cultivation and uses. In addition, there were many cases like that of Sloane, whose wealth came in part from Jamaican plantations that were based on enslaved labor. Often ignored in the past, botany's relationship to slavery is now being more carefully scrutinized along with other colonization practices.[10]

Keeping Order

Sloane had catalogues for each portion of his collection and wrote most of the entries and specimen labels himself, so he was personally involved in ordering his collection. One historian argues that these entries are an even more significant textual legacy than Sloane's volumes on Jamaica. The bigger the collection, the more a catalogue was needed. If the collection became disorganized or even destroyed, a catalogue continued to do its job of communicating to others what had existed at a particular time, what had been accomplished in the past.[11]

There are many ways to order plant specimens. One is alphabetically, but this ignores the similarities among species. If resemblances are used as the criteria, there are any number of bases on which to organize plants: for example, similar flowers or leaves or the presence or absence of woody tissue. Petiver grouped plants geographically, seeing that as the most significant factor. An alternative was listing the specimens in a ledger, perhaps in the order acquired. In any case, there had to be a relationship between the catalogue and the physical arrangement of the items themselves—otherwise, there was chaos. And often there was, especially after a collector's death. Even collaborators might not know all the ins and outs of the person's system, if indeed there was one. It was not uncommon for a new owner holding different priorities as to important information to decide that new labels were needed. The old labels, and perhaps vital or at least interesting data, were discarded, and reordering was frequently left unfinished, leading to more confusion later.

Unfortunately, Sloane's plant catalogues for his herbarium no longer exist. However, he had an alternate reference system based on the botanist John Ray's compendium of plants, completed in 1704. Next to the species entry in Ray, Sloane and his curators noted the herbarium volume and page where the specimen could be found, with the addition of species that were not in the text. Botanists still consult this reference today. Copies of Sloane's two volumes on Jamaica were used similarly. The latter were considered so significant that the names were updated by later botanists to keep them relevant for research. These resources indicate how important texts are as adjuncts to specimens, from simple lists and catalogues to entire libraries.[12]

The passion botanists brought to collecting plants they usually also brought to collecting information, as seen in the flow of correspondence among early modern botanists. Books and notes as well as specimens and drawings were shared. It is not a coincidence that botany burgeoned as printed books became more accessible. While many book collectors did not collect plants, those with herbaria and financial means, such as Aldrovandi and Sloane, were driven to create libraries and collections of botanical art. It would have been unthinkable for a botanical collector like Sloane not to have a reference library to support his herbarium and help make sense of it. He had the latest works by Plukenet, Petiver, and Ray as well as earlier classics like those of Pietro Andrea Mattioli and Caspar Bauhin, considered fundamental references. When the great classifier Carl Linnaeus began to publish, his works entered Sloane's library. The two also exchanged correspondence and specimens, with Linnaeus visiting London early in his career and studying Sloane's plants, though he was not impressed by their organization.[13]

Other Collections

Besides his herbarium, Sloane had a "Vegetable Substances" collection with twelve thousand small sealed boxes filled with seeds, fruits, and resins. Though this has not received much attention until recently, its value today rests in its history, extent, and rarity. There are many written references to botanists sharing seeds and other plant materials, but little of the physical material survives. Remarkably, about two-thirds of Sloane's items remain, as do three catalogues, which for most items list who sent the material and its use. Sloane's boxes provide a picture of what was considered valuable, including materials with medicinal uses. Collectors like Kaempfer forwarded seeds that might have been viable when they arrived. Some of these went to gardeners of Sloane's acquaintance who eagerly attempted to grow new finds.[14]

As time went on, other plant materials were also collected more systematically by botanists. These included wood samples, particularly as deforestation in Europe became a more serious problem. There was a search for species that would grow well and also provide good timber or other valuable materials, such as fruits or nuts, or be attractive additions to a large estate garden. A wood collection came to be called a xylarium (xylem is woody plant tissue) or even xylotheque, a play on the French word for library: *bibliothèque*. Some collectors took this term literally. One of the most notable was Carl Schildbach, who completed his around 1788 in Kassel,

Germany. He managed a menagerie for a German prince, giving him access to forests on the prince's estates. All 546 "volumes" are book-shaped boxes. Each spine is made of bark from the tree named in Latin and German on the spine's leather label, with the box built from different cuts of the species' wood. The interior is lined with blue paper and contains dried leaves, twigs, and fruit (fig. 5.2). The front edge has notes on samples of polished, dried, and burnt wood along with a fungus found on the species. Most xylotheques are less elaborate, simply book-shaped blocks of wood, often with labeled bark spines. More recent xylaria are composed of smaller samples, from which thin slices are sometimes removed for microscopic examination. These are useful in identifying suspected illegal shipments of timber from protected species.[15]

Other plant materials proved harder to store in ways that would conserve their characteristics. Eventually techniques were found for preserving mosses; these smaller, more simply constructed plants are often kept in packets rather than pasted onto sheets. Fungi presented an even greater challenge. Macrofungi, mushrooms, have high moisture levels and can shrink on drying to the point they become unidentifiable. Still, they are preserved as well as possible, often along with images of their living state and sometimes with "prints" of the spores released from the underside of the mushroom cap, an example of how botanists gradually found ways to document different materials. It should be noted that since the mid-twentieth century fungi are no longer considered to be in the plant kingdom; they are classified separately because of their very different composition and structure. However, their study is still considered part of botany.

For fleshy fruits, which do not preserve well, the solution was to submerge them in alcohol or other preservatives, and many herbaria have collections of jars. This system involves its own storage issues of weight and the necessity to top off the preservative fluid as it slowly evaporates. Delicate flowers like orchids can also be kept this way, while the flesh of cactus pads is scooped out and skins are dried and pasted on herbarium sheets—a delicate and often painful process due to cactus spines. Then there are boxes for large pine cones and six-foot-long palm fronds folded on themselves, as well as small wooden or plastic boxes for tiny aquatic plants

Figure 5.2. Interior of "book" for the Norway spruce (*Picea abies*) from the Carl Schildbach xylarium. (Photograph by Peter Mansfeld, Naturkundemuseum Kassel)

mounted on glass microscope slides. Often most daunting are fossil plants encased in rock, some weighing ounces, others hundreds of pounds. Curators have to be not only knowledgeable but ingenious in dealing with the difficulties they face in documenting plant life's variety. These specialized collections often get donated to herbaria that specialize in them because they have the facilities to properly handle them.

For difficult-to-preserve materials, drawings are particularly important adjuncts to specimens. But at many points in botanical history, three-dimensional solutions have also come into play. In eighteenth-century Italy and France, wax models of fruits and nuts and even of entire plants were created, along with the more famous wax figures of human anatomy. In the nineteenth century, papier-mâché was favored and large models with removable parts were used in education. Fungi were common subjects because the organisms were so difficult to preserve. In the twentieth century, William Dillon Weston, an expert on fungi, created glass models to make microscopic species easier for students to visualize.[16] And then there are the Harvard University glass flowers mentioned in chapter 1. While the latter have always been highly valued, many of the other plant models and educational posters popular in the nineteenth and early twentieth centuries were ignored or discarded. However, their value, especially as aesthetic objects and as artifacts documenting a past scientific era, has risen considerably in the twenty-first century. Many models and posters now decorate herbaria.

Value and Status

The Harvard glass flowers donated by the Ware family and other significant collections are accessible today thanks to wealthy patrons who had enough understanding of botany to support it and perhaps, in the process, become known for scientific sophistication. Some, like the Wares, were not as interested in science as those like Hans Sloane, who participated in inquiry. There are many permutations of the relationships among money, science, and status, and collections are important physical manifestations of these relationships.

While Sloane's herbarium volumes are bound in leather, this was obviously a working collection. He did not feel the need to have matching bindings or decorative pages; the specimens spoke for themselves. Because they came from many other collectors, the plants are preserved on paper of many different dimensions; the books range from the size of today's trade paperbacks to others with elephant-folio-sized pages. The quality of the paper also varies, as does the worth of the specimens and the information attached to them. Many pages have been annotated by later botanists—even with ballpoint pen, to the horror of one curator. Pencil is now the instrument of choice for adding information directly to any herbarium sheet. Leafing through a number of these books gives a sense of a botanical community stretching over space and time, intent on understanding the plant world. These volumes are still used by researchers to compare, for example, specimens collected today with those found in the same collection area more than three hundred years earlier.

Sloane saw his status as tied to the botanical quality of his collection, not to how his plants were presented. This was in contrast to collectors like Carl Linnaeus's patron George Clifford, a banker in international trade, who had his specimens emerging from printed vases with information written on elaborate labels (see fig. 3.4.) Here was a collector who wanted to be considered aesthetically as well as scientifically sophisticated. For herbaria that were given to patrons, the decoration could be even more elaborate. There is a bound herbarium purported to have been created by an Italian pharmacist for a physician, probably his patron.[7] It harkens back to the age of manuscripts with a title page decorated with floral borders and two watercolor scenes. The specimen pages have the plant names in calligraphy, sometimes with intricate flourishes in red, and the plants' stems are held down with colored silk ribbon (fig. 5.3). In keeping with their presentation, the specimens are of high quality, generous in size, and with minimal blemishes. There are many similar collections, often hidden away in libraries rather than in herbaria because they are considered more of cultural and aesthetic value than scientific.

For some collectors, the status a collection provided was its main value, a way to display both wealth and intellectual sophistication. But the value of any object is variable and so is what's meant by "value": scientific, monetary, sentimental, cultural, aesthetic. The fact Sloane bought so many

Figure 5.3. Specimens from a bound herbarium with elaborate calligraphy, attributed to Carlo Sembertini. (Oak Spring Garden Foundation, Upperville, VA)

collections suggests their monetary worth at that time. They provided significant equity for less wealthy botanists who depended on support from patrons. In the eighteenth and nineteenth centuries, many large herbaria were auctioned, often in a number of lots, resulting in collections being scattered, making it difficult to locate specimens mentioned in journals or correspondence.[18] As collecting became more institutionalized, private collections were less common, but even today some individuals have substantial and botanically valuable specimens that are often donated to herbaria.

A specimen's value can increase as its history lengthens, as it is moved from one collection to another or reexamined by fresh eyes, either as a scientific document or a cultural one: it is both. In 1906, Otis Mason, ethnology curator at the Smithsonian, wrote: "An ideal specimen is an object that has something to teach about humanity. . . . In the untaught mind it is a curiosity or monstrosity, and the more mystery there is about it, the better. But all such notions are far from the sciences of Anthropology. A good specimen is capable of telling more than one story. It may talk about race, development, geography, progress, skill, art, social life, or whisper of a spirit world."[19]

Recently, researchers at the Museum of New Zealand Te Papa Tongarewa analyzed DNA from six specimens of *Sophora* species collected in European botanical gardens around 1800. In the late nineteenth century, the museum's director asked colleagues in Britain to help him build the collection, and these were among thousands of sheets he received. But as often happens in herbaria, most of these donations remained unstudied because of the press of other work. The DNA tests revealed that one specimen was *Sophora toromiro*, native to Easter Island, Rapa Nui, but now extinct there, though it grows in several botanic gardens. It was thought that gardens began cultivating it only when seeds were collected after 1920. Yet here was evidence that the plant was growing in Germany in 1800; how could this be? The specimen could not reveal the answer, but archival records contained clues. The seeds may have been gathered during Captain Cook's second round-the-world voyage (1772–75) when the expedition stopped at Rapa Nui. The botanists on board were the first Europeans to collect specimens on the island, and one of them gathered seeds. He may very well have obtained some from *Sophora toromiro*, since it grew in thickets and was the only native shrub on the island. If the seeds were planted in the late 1770s, then the shrub would have been established enough to yield cuttings in 1800.[20]

It may not be possible to find the backstory of every specimen as for this *Sophora,* but investigations of individual items like this increase the value of an entire collection. This is what makes old herbarium specimens wonderful: their stories are manifold and often intriguingly hidden. Hans Sloane's herbarium is massive and is only beginning to be mined for its historical value, and there are many smaller but nonetheless rich collections in herbaria and libraries, large and small.[21] Some specimens are still lying undiscovered, waiting to be found, conserved, and analyzed: in other words, they await the needed intervention of curators.

Curation

Curation, the care and study of collections, is important to maintaining items and revealing their stories. The field has grown more sophisticated over time, but there are still judgments made by one generation of custodians that are questioned or reviled by the next—about everything from the glue used to paste down specimens to the ordering system to the decision to dispose of "worthless" items. This adjective's meaning changes with time. Yes, some specimens should be culled, perhaps because they are so damaged by fungi or insects that there is not enough left to identify the plants. However, the labels might be worth keeping if they have information on the plants and where they were found. It's not always easy to tell what has long-term worth, and that's a problem for a curator: to evaluate an item in terms of the past, present, and future. The Sloane collections that exist today are a portion of what Sloane himself amassed. After his death, they had to be stored for some time before the British Museum was constructed to house it. Even then, many items remained in storage, and some deteriorated badly, especially the zoological specimens, many of which had become little more than dust. Few of these are extant, while the plants, most in bound volumes, fared much better.

As items were added to its various collections, the British Museum soon grew much larger. By the nineteenth century, the natural history collections could no longer be fully accommodated. The solution was construction of the Natural History Museum, London, which opened in 1881. It too became inadequate and additions were built, including the Darwin Centre, which opened in 2009 with a climate-controlled room and specially built shelves for each of the 265 Sloane Herbarium volumes. Meanwhile, the library in the British Museum also outgrew its space, and the British Library was built at a separate location. It contains Sloane's books and manuscripts, though his artworks, coins, and artifacts remain in the museum.

These shifts mean that the Sloane collection, originally unified in his home, is now in three different locations spread across London. If a researcher wants to examine Paul Hermann specimens and manuscripts, visits to both the Natural History Museum and the British Library are needed. Looking at botanical art may require a trip to the British Museum. Many insoluble problems in curating a large collection force compromises between order and ease of access. This is why digitization of collections, discussed in chapter 15, is attractive to curators and researchers as a way to allow the juxtaposition of different materials that could enhance the understanding of each: specimens, notebooks, illustrations. This ability to make connections would be particularly useful in the work of what is called "decolonizing collections," that is, finding the stories of the indigenous and enslaved people whose roles were so important in their acquisitions but were unacknowledged in the past.[22]

SIX

Linnaeus and Classification

THE ENLIGHTENMENT'S GUIDING principles of reason and prog-
ress were among the influences driving natural history in the eighteenth century. A major force generating a passion for nature was a Swedish botanist
with a thirst for order, Carl Linnaeus. Interested in plants from an early age,
he came to a vision of that order as a medical student at Uppsala University.
He had transferred there from Lund University, where one of his professors
had introduced him to the herbarium as a way to document plants he encountered. Linnaeus was never much of an artist, making the plant itself an
even more essential reference. He would later write that there was no substitute for an herbarium, that it was the crucial source of information about a
plant.[1] His collection would ultimately become the most botanically significant in existence because of his role in naming species.

When Linnaeus arrived in Uppsala in 1728, he contacted those at the
university who shared his interest in plants. Botanists there had studied the
seventeenth-century plant collection of Joachim Burser, who had based his
collection on the work of Caspar Bauhin, author of one of the best botanical
references of the sixteenth century. These specimens would be valuable to
Linnaeus years later, suggesting their enduring value. Linnaeus was men-
tored by Olof Celsius, a professor of theology and also a serious amateur
botanist. They met by chance in early 1729 at Uppsala's botanic garden, and
Celsius was impressed by the younger man's knowledge of classification and
by his herbarium, which already numbered six hundred specimens. Celsius
offered Linnaeus room and board in return for help in finishing a book on

plants mentioned in the Bible. At about the same time, Linnaeus became friendly with a fellow student and naturalist, Peter Artedi, who was also searching for a way to organize what he was learning about the living world. Instead of competing, they decided to divide up the animal and plant kingdoms between them, an audacious project for young men in their twenties, and together they continued it off and on until 1734, when each left Uppsala.[2]

Travels

Among Linnaeus's notable learning experiences was a 1732 government-sponsored trip to Lapland in northern Scandinavia to survey its natural history in search of economically attractive resources for development. Linnaeus improved his collecting skills and returned not only with plants but with animal skins, insects, and mineral samples. The most significant results from the trip were his journal, which presented observations on the natural history and people he encountered, and a flora of Lapland featuring descriptions of plants he had found there.[3] This was Linnaeus's only long collecting trip; he made several shorter ones after he became a professor, but those were years in the future. In order to gain more research experience, he set out in 1735 for the Netherlands, where the botanical resources—gardens, herbaria, and botanical libraries—were richer than in Sweden. He brought with him an amazing set of manuscripts he had produced in Uppsala; these amounted to the essentials of the classification system he spent his life elaborating.

Among these writings was one laying out Linnaeus's approach to classifying plants. While he is justly seen as the originator of many taxonomic innovations, as with most cases of scientific genius, his work built on that of his forerunners. John Ray in England and Joseph Pitton de Tournefort in France had written books describing thousands of species and presenting ways of ordering these to emphasize the relationships among them.[4] Ray explicitly delineated the definition of a species, the idea that seeds of one generation produce the next generation of like individuals. He saw the species as the basic unit of taxonomy and described species based on traits that remained stable across individuals and generations.

Tournefort agreed with Ray on what a species was and refined Ray's concept of the genus, a group of species that share major characteristics. When many species are examined and compared, it becomes apparent that they can be sorted into natural groups, that genera exist in nature, not just in the minds of botanists. For example, pitcher plants in the *Sarracenia* genus have similarly structured flowers and also have leaves that fold around themselves longitudinally to form vase-like structures. These fill with fluid so insects that slip down the pitcher's sides drown and are digested. There are other pitcher plants of the genus *Nepenthes* in which the flower and the pitcher structures are completely different: a tendril swells at the end and forms a swollen sack that fills with fluid to trap insects and even small animals. Genus and species concepts seem relatively obvious today, but at the time the ideas were still being elaborated, their meanings just emerging. Ray, Tournefort, and others were documenting information on these concepts. They were building on the improved definitions of plant anatomy botanists were publishing, such as Sébastien Vaillant's descriptions of reproductive organs—the male stamen with anthers containing pollen and the female pistil with egg-containing ovules (fig. 6.1).[5]

While there was some agreement among botanists about what constituted a species and a genus, higher levels of organization were debatable. As Tournefort had focused on flowers in his classification system, so did Linnaeus, but with a difference. Tournefort looked at the flower as a whole and also included other characteristics in his organizational hierarchy. He and Ray were aiming for what is called a natural system, attempting to group genera into broader categories—families and then classes—that shared a number of traits. This was a difficult task, and Tournefort and Ray ended up with similar but still differing results. While accepting their views on the existence of species and genera in nature, Linnaeus was looking for something simpler for higher levels of organization, a method not based on a range of traits. He devised an artificial system, also centered on the flower, but focusing only on the male and female structures: the stamens and pistils, respectively. He divided plants into twenty-four classes based on the number and placement of the male stamens. These classes were subdivided based on the number of female organs, the pistils. Plants could be sorted simply by count-

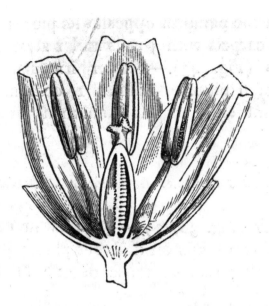

Figure 6.1. Cross section through a tulip flower, adapted from Joseph Dalton Hooker's *Botany*. Petals surround three male stamens topped by pollen-carrying anthers; the central female pistil contains ovules that will develop into seeds. (University of California Libraries)

ing structures that were usually visible to the naked eye, though some dissection of a flower might be necessary and use of a hand lens desirable.[6]

Early in his stay in the Netherlands, Linnaeus reconnected with Peter Artedi, who was working on his fish classification system. However, Artedi accidentally drowned in one of Amsterdam's canals, leaving Linnaeus to see his friend's ichthyological manuscript published and to complete their plan to classify all plants and animals.[7] This could not be accomplished without contacts and wide correspondence. Many individuals provided not only plant material but information on animals and minerals as well, though the focus here will be on plants.

Linnaeus definitely honed his networking skills in the Netherlands. He began by showing the outline of his classification scheme, *Systema Naturae*, to Johann Friedrich Gronovius, a botanist with an impressive herbarium that included plants from collectors worldwide. Gronovius was knowledgeable enough to appreciate the importance of Linnaeus's manuscript and had it published at his own expense. He also aided the young botanist by sharing specimens he had received from British collector John Clayton, who settled in Virginia.[8] Gronovius had Clayton's work published and allowed Linnaeus to examine the specimens and the manuscript, giving the young botanist his first opportunity to study North American plants.

Gronovius also introduced Linnaeus to Herman Boerhaave, who had been professor of medicine and botany at Leiden University and head of the botanic garden there, the one founded by Carolus Clusius. Boerhaave had an extensive herbarium with the sheets kept loose so that he had the flexibility of reorganizing specimens. He had published Sébastien Vaillant's work after the latter's death, so he was well aware of the latest research on floral reproductive organs. Now Vaillant's collection was available to Linnaeus, and Boerhaave also gave him a letter of introduction to Johannes Burman at the botanic garden in Amsterdam. He was working on the collections of Paul Hermann from Sri Lanka, opening a different portion of the plant world to Linnaeus. Burman urged Linnaeus to visit yet another member of the Dutch botanical network, George Clifford, who became Linnaeus's most important contact in the Netherlands.

Hartekamp and Beyond

Clifford's garden at his Hartekamp estate was filled with exotic plants from around the world. He was a banker and a director of the Dutch East India Company, a major trading enterprise, giving him access to the latest species arriving from South Africa, India, and the Far East. To accommodate them, Clifford had four hothouses, one each for plants from Europe, Africa, Asia, and the Americas, a living collection backed by an herbarium and botanical library—everything Linnaeus needed to learn about the world's flora.[9] Making the situation ideal was that Clifford invited Linnaeus to live at Hartekamp in the midst of this botanical wealth and write a catalogue with descriptions of the plants in his collection. Linnaeus remained there for two years in what amounted to a postgraduate education in botany.

Clifford also financed a trip to England so Linnaeus could study collections there, extending his knowledge still further. Each herbarium has a unique set of specimens, depending upon the interests of its owner and also on accidents of luck as to what materials are available. Hans Sloane's massive collection was at the top of Linnaeus's list of must-sees, and he secured a letter of introduction from Boerhaave, who unfortunately went a little too far in his praise. He put Linnaeus on a par with Sloane, describing them as "a pair of men whose equal is hardly to be found in all the world."[10] This did not sit well with the owner of the largest herbarium of the time. Sloane did not pay much attention to young Linnaeus, who later described Sloane's herbarium as disorganized. At the time, Sloane was incorporating James Petiver's large collection into his own.

Meanwhile, Philip Miller, head gardener at the Chelsea Physic Garden, was dismayed that Linnaeus spent more time studying Miller's herbarium than examining the broad selection of exotic plants growing at Chelsea. However, he gave Linnaeus an array of plants to take back to Clifford as well as herbarium specimens that a British collector had brought from Central America, revealing still another flora to Linnaeus.[11] A Quaker merchant, Peter Collinson, was also on Linnaeus's list of contacts in London because he grew plants from around the world. He got along well with Linnaeus and gave him a package of specimens and seeds for Clifford. Collinson continued to correspond with Linnaeus over the years, sending him plants and connecting him to other botanists and plant collectors.

Another key encounter facilitated by Clifford occurred before Linnaeus's British trip. Georg Ehret was a young artist who specialized in botanical watercolors. He began as a gardener in his native Germany, but soon came to the attention of his patrons for his artistic skills. He painted many illustrations for a German botanist who trained Ehret in presenting plants, and particularly flowers, in enough detail to make species clearly identifiable. Lack of affluent patrons drove Ehret to England, where he produced watercolors for avid gardeners seeking to memorialize their collections, Collinson among them. Then Ehret decided to try his luck in the Netherlands and visited Hartekamp, where Clifford was impressed with his work and commissioned him to create illustrations for the catalogue Linnaeus was writing. Ehret remained at Hartekamp for a month, working alongside Linnaeus, engaging in the four-eyed sight that resulted in accurate and aesthetically satisfying botanical art. Later Linnaeus claimed that he taught Ehret the proper way to draw plants with scientific accuracy, while Ehret asserted he had learned this in Germany. In either case, Ehret's output was stunning.

Besides paintings of individual species, Ehret also produced a masterful illustration of Linnaeus's classification scheme, presenting the sexual organs for each of the twenty-four classes, that is, groups of genera. It's an amazing chart that rewards careful examination (fig. 6.2). The first class, monandria, is composed of species with one male stamen and is illustrated in the upper left-hand corner along with one female pistil (A); the next class, diandria, has two stamens shown with a single pistil (B); and so on. The twenty-fourth class, cryptogamia, consists of all plants that do not have obvious sex organs (Z). This is a class in which the artificial nature of Linnaeus's system is most apparent because it includes flowering plants like the figure pictured in Z along with mosses and algae. The most ingenious drawings are to the left of this, representing species with separate male and female flowers on the same individual plant (V), on different individuals (X), and finally where a single plant has flowers with one or the other sex, as well as flowers with both sets of organs (Y). As these three classes indicate, plants are complicated and diverse, and Ehret clearly and economically presented distinguishing features.[12]

While living on Clifford's estate, Linnaeus completed the manuscript for *Hortus Cliffortianus* as well as a short book on the banana, the text revealing his abilities in botanical explanation. It describes a banana tree that had been growing in Clifford's greenhouse for some time; Linnaeus was able to coax it to bloom. This was quite a feat at the time, when many tropical exotics might be kept alive in hothouses but did not necessarily thrive and flower.[13] Linnaeus proved he could work with living plants as well as herbarium specimens. Later, in Sweden, he created a thriving garden with its own greenhouse so his students could see a variety of species in bloom.

While still in the Netherlands Linnaeus also published a number of brief works laying out the essentials of his classification system. There was

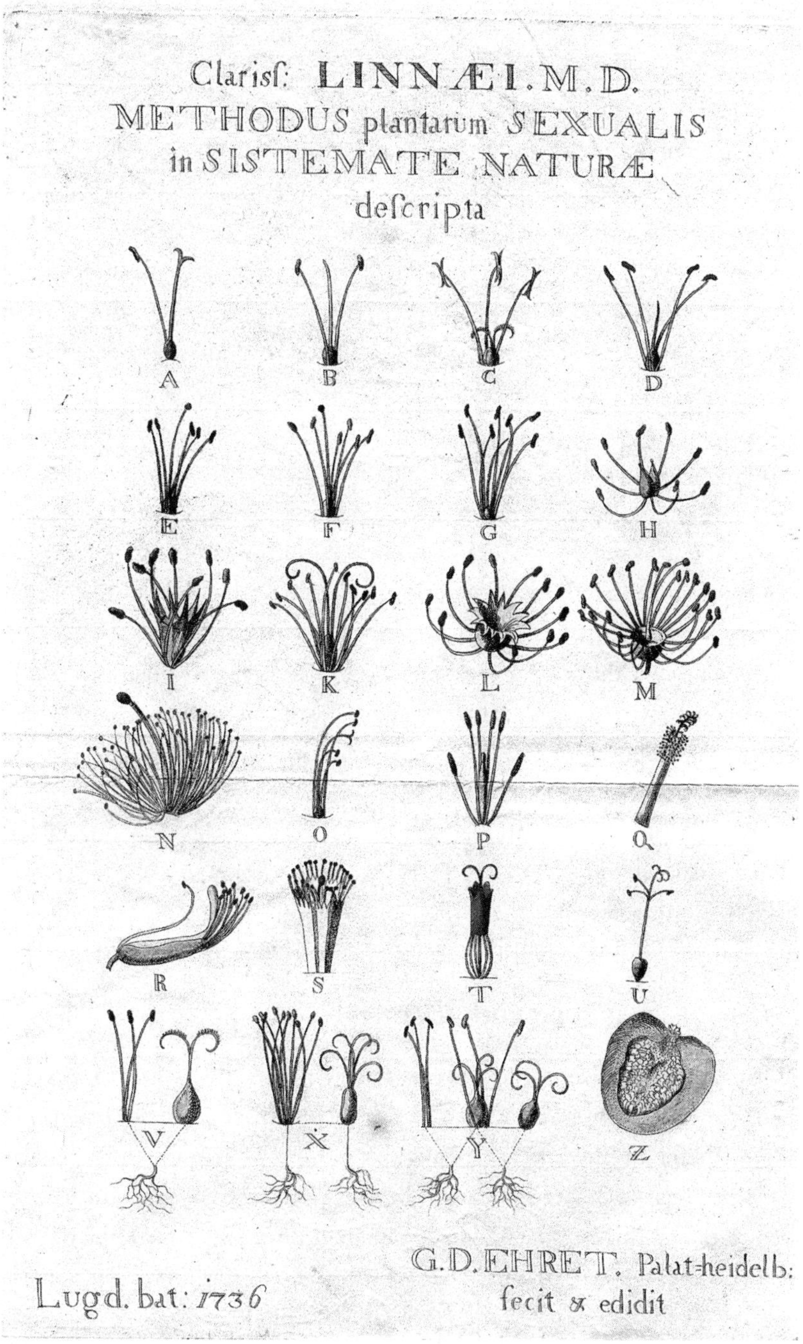

Clarisſ: LINNÆI. M. D.
METHODUS plantarum SEXUALIS
in SISTEMATE NATURÆ
deſcripta
A
B
C
D
E
F
G
H
I
K
L
M
N
O
P
Q
R
S
T
U
V
X
Y
Z
G. D. EHRET. Palat=heidelb:
fecit & edidit
Lugd. bat: 1736

Figure 6.2. Illustration by Georg Ehret of Carl Linnaeus's plant classification system.

Systema Naturae, mentioned earlier, in which, thinking big, he created a basic classification system for animals, plants, and minerals. In *Fundamenta Botanica,* Linnaeus set forth the botanical vocabulary and principles he used in organizing the plant world. *Classes Plantarum* described his twenty-four classes based on plants' sexual structures, and *Genera Plantarum* outlined the 935 genera into which he organized all plant species.[14] Linnaeus dealt with all the plants, both living and preserved, that he had encountered first in Sweden and then in his travels. Most of the genera and species had come from works of Ray, Tournefort, and Vaillant.

After two years at Hartekamp, Linnaeus stayed in Leiden for several months working with yet another Dutch botanist, Adriaan van Royen, director of the botanical garden there. Before heading back to Sweden, Linnaeus spent a month in Paris at the king's garden, home to many elite botanists, including Bernard de Jussieu, who allowed him to study Tournefort's herbarium, the basis for the Frenchman's classification scheme. Linnaeus also met the botanical artist Claude Aubriet, who had illustrated Tournefort's work and who showed Linnaeus the impressive assemblage of botanical paintings done over years at the garden. Through this experience in addition to his time with Ehret, Linnaeus developed a sense of how the best botanical art presented plants. Later he would rely heavily on illustrations in describing species that he did not have access to as living or preserved specimens.

Teaching and Encouraging Collection

With all this botanical work, it is easy to lose sight of the fact that Linnaeus had trained as a physician. When he returned to Sweden he began practicing medicine in Stockholm, having married the woman to whom he had become engaged before his travels. But his heart was not in medical practice, and after three years, he finally obtained the position he wanted: professor of botany at Uppsala. There he had access to a botanical garden, which was in poor condition but which he would improve over the years, even adding a hothouse. Adjacent to the garden was the professor's house, so he had a living collection as well as room for his growing herbarium.

By this time Linnaeus had seen many ways to organize and preserve specimens. Some were bound into volumes, like the Burser and Sloane collections were. This method preserved the specimens well, but made it difficult to re-sort them, so Sloane and his curators had to keep track of specimens by annotating a copy of John Ray's description of plant species. Instead, Linnaeus left his specimens as loose sheets and had a special cabinet built with twenty-four shelves, one for each of his plant classes (fig. 6.3). The shelves were not evenly spaced since some classes were larger than others. With this setup he could readily find a specimen and refile it elsewhere if necessary. After this, most herbaria were left unbound as reclassification became a continuing process extending to the present day. It became standard to store the sheets in folders to protect them, and in some herbaria the folders were then put into boxes to be stored on shelves.

At Uppsala, Linnaeus was an energetic teacher, researcher, and exponent of the economic importance of plants. Plant exploration had brought many useful plants back to Europe, including tomatoes and potatoes, which were slowly being introduced into the European diet; tobacco, coffee, and tea also became more and more popular. One reason Linnaeus was interested in new species, besides adding them to his ever-growing classification system, was to discover those that would thrive in Sweden, improving the country's agriculture and even its industry. He sought tea plants and palms from Asia and wanted to grow mulberry trees as a way to begin silk production since silk worms thrived on its leaves. Early in his career at Uppsala, Linnaeus arranged for one of his best students, Pehr Kalm, to receive funds for travel to North America in search of good botanical possibilities. Canada seemed a likely place to look since its latitude was similar to Sweden's. However, Linnaeus's connections meant that Kalm first landed further south, in Philadelphia, close to a Swedish colony that was founded in the 1630s and home to Benjamin Franklin, a friend of Peter Collinson, who wrote a letter of introduction for Kalm at Linnaeus's request.

Collinson knew Franklin would connect Kalm with John Bartram, a Philadelphia naturalist and nurseryman who had already begun a long-term correspondence with Collinson. Over about a thirty-five-year period, Bartram sent a large array of specimens, seeds, and seedlings to Collinson, who

Figure 6.3. Linnaeus's specimen cabinet at Hammarby, his farm outside Uppsala. Alongside the cubicles, the doors are marked with roman numerals for the twenty-four classes. The faded rectangles to the right are prints of Georg Ehret illustrations that Linnaeus had pasted to the walls of his bedroom so he could enjoy looking at them. (Gina Douglas HonFLS)

in turn distributed them to leading gardeners of the day keen to have the latest finds from North America. Collinson had even recommended Bartram to Linnaeus as a good source of New World plants. Kalm stayed at Bartram's farm for several days, studying his herbarium and the plants in his garden. Kalm eventually headed north, visiting Cadwallader Colden, a New York farmer who had contact with Bartram, Collinson, and Linnaeus, using the glue of correspondence to keep in touch. Colden had taught Linnaean classification to his daughter Jane, and it may have been Kalm who encouraged her to write descriptions of the plants she collected, pressed, and drew. Parts of her manuscript were finally published in 1963.[15]

Leaving Colden, Kalm went on to Canada, traveling as far as Montreal and Quebec. He wrote Linnaeus of his progress, collected specimens along the way, and packed up seeds for Sweden. He returned to Philadelphia in the winter, and the next spring headed west as far as Pittsburgh and then up to Niagara Falls. The journal Kalm later published carefully described observations of the plants and animals he found and also the hardships he endured in sparsely settled areas. Two collecting seasons were all his stipend would allow, so in February 1751 he returned to Sweden with a new bride and a collection that pleased his professor. Kalm grew the seeds he brought back, including North American mulberries, but many plants did not thrive and none would ever become economically viable in Sweden. Even the mulberry trees eventually died. Acclimatizing plants was not an easy task, and at best it took years of trial and error to achieve success. There were limits to where a plant could grow outside its native range. Linnaeus sent other students and protégés on voyages of discovery, but in the end there was little economic benefit to Sweden from his quest. However, they did send back a wealth of new species that Linnaeus added to later editions of his books on plant species and genera.[16]

Nomenclature and the Value of Linnaeus's Collection

As Linnaeus described more and more new plants, he encountered a problem that plagued all botanists up to that time: giving a new species a name to distinguish it from all others. Latin was still the language of botany at the time; all Linnaeus's works were published in Latin, including all the plant names. As more species were described, the names became longer, especially in cases where there was a large group of plants with similar properties. Some names were six or more words long, making them difficult to remember. By the mid-1740s, Linnaeus had devised an informal way to deal with the problem in his notes, giving each species in a genus a one-word "nickname"—or in Latin, a *nomen triviale*—to designate it, for example, one tulip species was called *Tulipa gesneriana* in honor of Conrad Gessner to shorten *Tulipa flore erecto, foliis ovato-lanceolatis*. The binomial was accompanied by a brief description with the plant's key distinguishing traits. At first the practice was just for his own convenience, but then he used it more formally in the index to his publication on the plants of Öland and Gotland, islands off the coast of Sweden where Linnaeus spent a summer collecting.[17] Later, it became standard for him, and others soon took up the practice.

What Linnaeus accomplished with this method was to separate the name from the species description, so the name did not have to do the work of both. It is ironic that binomial nomenclature, which Linnaeus originally termed trivial, would be a lasting contribution to botany—in fact to all biology, because he also published animal names as well, and the practice was extended to the microscopic world when bacteria were discovered. On the other hand, his twenty-four-class taxonomic scheme for plants, while it became popular and widely used, was eventually abandoned as others devised less artificial, more natural systems based on several traits.

Linnaeus's work became most widely referenced in his opus *Species Plantarum* of 1753, in which he attempted to describe all known plant species, over seven thousand at that time. *Species Plantarum* is an extremely important work in the history of botany: all plant names currently in use date from its publication or later.[18] Earlier names are deemed invalid. This is one reason why until recently most early herbaria, if they had not been consulted by Linnaeus, were considered of little value to present-day botanists, because the old names were no longer used.

The fact that pre-Linnaean collections could be important historical documents has only recently been widely recognized, leading to work, for example, on the Middle East specimens of Leonhard Rauwolf that document not only the native plants of the area but the cultivated ones as well. On the other hand, any specimen that Linnaeus set eyes on, that he might have used in writing his species descriptions, is considered extremely valuable. Cases where he never saw the living plant or even a specimen but where he instead had to rely on illustrations give them special worth. The reason will become clearer in chapter 11, which covers how botany later became more institutionalized and organized. Botanists became more convinced that plants themselves must be the final arbiters of their characteristics.

After 1753, Linnaeus continued describing new species, publishing several updated versions of *Species Plantarum*. As he had arranged for Pehr Kalm to gather specimens in North America, he sent out those he termed "apostles" to every continent but Australia, which was not well explored during his lifetime. There were about twenty of these men, several of whom never returned from their journeys. Peter Forsskål died in Yemen after collecting there and earlier in Egypt. His collections are particularly important because he recorded Arabic as well as Latin plant names. Others, like Pehr Kalm and Carl Thunberg, returned and had long careers in botany. Thunberg collected in South Africa, Japan, Ceylon, and Indonesia; then he became professor of botany at Uppsala and held the chair for forty-four years.[19] While there are few Linnaeus specimens at that university, Thunberg's collection is preserved there along with the earlier herbarium of Joachim Burser, which was studied by Linnaeus from his school days at Uppsala.

At his death, Linnaeus's natural history collections, which included plant, animal, and mineral materials, were already regarded as valuable. His son, who succeeded him as professor of botany at the University of Uppsala, took steps to preserve them. When Carl Linnaeus the Younger died five years later, Linnaeus's wife sold the collection in order to secure her financial future. By this time an herbarium could be a significant portion of a botanist's estate as well as a scientific legacy. There were few institutional herbaria; a collection usually belonged to the individual who had amassed it.

The British botanist James Edward Smith was able to purchase not only the specimens but Linnaeus's manuscripts, library, and even three of his custom-built herbarium cabinets.[20] All remained in Smith's possession until his death, when his heirs sold Linnaeus's material, along with Smith's own collection, to the Linnean Society of London. Smith and his colleagues had founded the organization after the original purchase, and it has done an excellent job curating this treasure. In the 1970s the society had a special underground vault built for the collection in its headquarters. This was after the materials were moved out of London during World War II, when much of the massive Berlin herbarium had been destroyed, highlighting the vulnerability of such collections.

By the end of the eighteenth century, the herbarium had become an essential element in botanical science, thanks to many botanists, including Linnaeus. While earlier some botanists such as John Ray used herbaria but did not think it necessary to create large collections, these became de rigueur for serious botanists. The need to follow the example of the greatest botanist of the century was one reason, but another was the result of Linnaeus's classification system. He chose a limited number of traits in describing species, ones that were usually visible in specimens, though identification might involve rehydrating flowers to recover some of their three-dimensional form.

As Linnaeus explains in his *Philosophia Botanica,* he based his classifications on four elements: the number, position, size, and arrangement of flower parts.[21] These strictures made it easier to describe species in words; *Species Plantarum* has no illustrations. Where illustrations were needed, they could be pared down, as they were to an extreme in Ehret's diagram of the Linnaean scheme. In counting stamens and pistils, Linnaeus was influenced by the mathematization of science that went back to Galileo and Newton. He wanted to bring order to living things, as these men had to the physical world, by simplifying the multiplicity of individual observations. In the process, he opened up botany to a broader audience by making classification easier, basing it on readily identified characteristics. Students and amateurs could name plants with less effort, making botany a more common hobby. It became more common as well for plant collectors to explore territories with unfamiliar species.

SEVEN

Botanical Exploration

IN THE EIGHTEENTH CENTURY, expeditions became more sophisticated thanks to new navigational aids, better maps, and the experience of earlier travelers. In terms of botany, Linnaeus's classification system made it easier for plant hunters to identify what they found and to appreciate new species when they came upon them. Governments grew more interested in systematically searching for useful plants. In its Latin American colonies, Spain at first exploited the land primarily for raw materials, particularly precious metals and other minerals. However, in the latter part of the eighteenth century, the Spanish king, urged by the head of Madrid's Royal Botanical Gardens, among others, sent out expeditions focused on plant collection as well as mapping and geographic studies aimed at discovering economically useful species. Each was composed of a team of naturalists, artists, and collectors. These expeditions were of longer duration than those mounted by other nations, in part because they worked in areas with already well-developed Spanish colonial infrastructures. Despite this, few economic benefits resulted from the enterprises.[1]

Though a great many plants were collected and drawings produced, little was published on the collections. There were several problems, from loss of specimens at sea to political upheaval to the lack of driven botanists like James Petiver to get species descriptions into print. The team headed by José Celestino Mutis in New Granada, which covered a good deal of northern South America, produced some sixty-seven hundred vivid botanical illustrations over a twenty-five-year period, but during that time, Mutis and his collaborators wrote only about five hundred plant descriptions. They sent twenty thousand specimens back to Spain, but many were not studied until the twentieth century and some have yet to be. Mutis considered drawings more important than specimens.[2] He created a program for Spaniards to train indigenous artists in botanical drawing. Their watercolors were accurate portrayals of the plants, including enlargements of flower parts, but their style was distinctly different from European botanical illustrations. Mutis's artists used saturated colors, filled the available space with relatively flat plant forms, and emphasized sinuous features (fig. 7.1).

The botanists Martín Sessé y Lacasta and José Mariano Mociño, accompanied by botanical artists, traveled in Mexico and parts of Central America from 1787 to 1803. Their plan was to complete the work that Francisco Hernández had begun two hundred years earlier. When they returned to Spain, the two began writing up their results with the aid of specimens and over twenty-two hundred illustrations. Sessé died in 1808 at about the same time the French invaded Spain. Mociño received support from the French to continue his work, but when the French were driven out, he was condemned as a collaborator and fled to France to avoid arrest. At Montpellier he met the Swiss botanist Augustin Pyramus de Candolle, who was writing a massive botanical compendium and was interested in the species depicted in drawings Mociño brought with him. The Spaniard allowed de Candolle to copy many of them and publish descriptions.[3]

Eventually Mociño returned to Spain, but he never published any results, and specimens from the expedition remained unstudied in Spain. In the mid-twentieth century they were sent to the Field Museum herbarium in Chicago to be photographed. Using these photographic resources and other information, Rogers McVaugh and his colleagues eventually produced several volumes on the expedition's plant collections, the last appearing in 2000.[4] While the fate of Sessé and Mociño's discoveries was not a typical outcome, it was common for only a portion of an expedition's plants to be described because of the many steps involved and the opportunity for missteps from the initial finding to publication. Frequently the collector was not the describer and, as is obvious here, sometimes there were numerous middlemen, as well as difficulties ranging from the death of experts to lack of financial backing to political disruption.

Figure 7.1. Watercolor of serápias (*Elleanthus magnicallosus*) from the Royal Botanical Expedition of the New Kingdom of Granada (1783–1816), directed by José Celestino Mutis. (© Real Jardín Botánico-CSIC)

French Exploration

The French and British were also mounting large expeditions in the eighteenth century, always with an eye on each other's moves, since they warred several times, jockeying to maintain and enlarge their colonial influence and their sway in Europe. Scientific and governmental networks became increasingly entwined as naval officers were trained in natural history and drawing so they could gather specimens and document what they saw. Ships became floating labs for preserving and even identifying collections with the aid of Linnaeus's system.[5]

In 1766, Louis Antoine de Bougainville, head of a French around-the-world expedition, appointed Philibert Commerson as naturalist/physician after he submitted a seventeen-page description of what he would accomplish. Commerson's charge was to find medicinal plants. When the Bougainville expedition reached Mauritius on the return voyage, Commerson decided to remain there because he was fascinated by the plants on the island and on neighboring Madagascar. On Mauritius, he worked at a botanical garden established by another French botanist, Pierre Poivre, who had traveled to the Dutch-held Molucca Islands and secretly brought nutmeg and clove plants to the Mauritius garden, thus breaking the Netherlands' monopoly. This was one of many botanical espionage missions sent out to ease supply bottlenecks. Commerson assisted Poivre while still searching for new plants until he died at age forty-five. Though his herbarium, with about three thousand new species and his notes, reached Paris, they were not studied systematically. So he is not always given credit for his finds, a common problem for collectors, especially those who do not describe the plants. There is a great deal involved in studying a collection once it is back in the collector's home country. It has to be organized into families, and the specimens mounted and labeled. Ideally, experts in that plant family will study these specimens and determine if they are distinct from known species or even known genera. This requires studying other specimens and also written descriptions in the literature. It is a lengthy process, and a daunting one if there are piles of plants, resulting in many collections being only partially studied.[6]

Another major French expedition sailed in 1785 under the command of Jean-François de La Pérouse. The enterprise was well staffed scientifically. La Pérouse sent back lengthy reports and materials from their stops. However, the naturalists complained of little time to collect, a problem on many expeditions. Voyages had multiple aims and natural history was rarely at the fore. La Pérouse promised to remedy this after they reached Australia, which they did in early 1788. They made amicable contact with the British at their new colony near present-day Sydney. The French left in early March for the Solomon Islands, planning to go on to explore the southern and western Australia coasts, but they were never heard from again.

In 1791, the French outfitted a new naval mission to learn the La Pérouse expedition's fate and continue its work; included was Jacques Labillardière, a botanist who had experience collecting in the Near East. The expedition stopped at Cape Town, South Africa, but through a series of mishaps the large collection made there never arrived in France. They also never discovered what happened to La Pérouse; evidence of his ship's wreckage on one of the Solomon Islands was not found until 1826. While stopping in Australia, New Zealand, Tasmania, and the East Indies, Labillardière collected about ten thousand specimens as well as seventy tubs of live plants and six hundred kinds of seeds.[7] These varied materials suggest the multiple needs collections served: specimens for documentation and description of new species as well as seeds and seedlings to test species' possible uses as food, garden plants, timber, fiber, dyes, and medicines. Transporting live plants, though difficult, was attempted when seeds were hard to come by or difficult to germinate.

European politics upset plans for the safe transport of these collections back to France. When the expedition members arrived in Dutch-held Java in 1793, they learned that the French king had been executed, provoking war between France and the Netherlands. Officers and naturalists on board were arrested by Dutch officials, but were treated differently depending upon their political leanings. Royalists, including Labillardière, were held, while republicans were allowed to sail home. Though he hid his journals and collection notes, the Dutch seized Labillardière's specimens, sending them with the French who had been released. In a later twist, the British boarded their ship and confiscated the collection.

Labillardière did not return to France until 1796. By that time, twenty-eight crates with his specimens were in England, where the French court was living in exile, welcomed by a sympathetic monarchy. The French royals received the collection because Louis XVI had been king when the expedition sailed. The exiles offered Britain's Queen Charlotte, an amateur botanist with her own herbarium, an opportunity to select specimens. However, Labillardière petitioned Joseph Banks, as a fellow botanist and confidant of the British king, to return the specimens in the name of science. Banks considered science above politics and was attempting to maintain contact with French scientists despite the repeated hostilities between the countries. In a noteworthy act of magnanimity and fairness in herbarium history, Banks returned the collection without even looking through it: a feat of self-control for a keen collector with his own impressive herbarium. A little later, Napoleon Bonaparte took a different view and saw herbaria as spoils of war. When he marched into Italy he sent the head government gardener from Paris to select choice specimens from Italian collections, and did something similar when he moved into Spain. This is in addition to the collections made by French botanists in Egypt during Napoleon's reign there.[8]

Since Labillardière had held on to his notes and journals, he was able to make full use of his specimens and publish his results, with illustrations by the renowned botanical artist Pierre-Joseph Redouté. His descriptions were thorough, including indigenous names and uses. His herbarium was one of the last collections created with government support that did not end up in the national collection in Paris. Labillardière, who fought hard to recover his specimens, also fought to keep them. They were purchased from his estate by the botanist Philip Webb, who gave his entire herbarium to the Natural History Museum in Florence.[9]

Britain and Global Botany

Joseph Banks, hero of the Labillardière affair, was a wealthy Englishman who early in life developed a passion for plants. He attended Oxford and then settled in London, where he began visiting the British Museum, home to Hans Sloane's herbarium. There he met Daniel Solander, one of Carl Linnaeus's former students, who was invited to Britain by Peter Collinson, a British Museum director and avid gardener, to make his mentor's classification system better known there. Early in his stay Solander updated Collinson's specimens with binomial nomenclature and did the same for Collinson's friend, the Duchess of Portland, who had the collecting bug almost on a par with Sloane's. Concerned about the future of Sloane's herbarium at the British Museum, Collinson thought Solander was right for the job of assigning Linnaean names to the specimens. He suggested this to Lord Bute, the prime minister and adviser to young King George III, who made the appointment. That this would be a matter of importance to Britain's ruler illustrates the place of botany in British society and culture at the time.

Banks learned much from Solander, who updated the names on Sloane's specimens without removing the original labels, as had often been done earlier. This was an important step forward in preserving a specimen's history and a sign of maturing herbarium protocols. Still used, this practice explains why some specimens are littered with paper slips noting name changes or reaffirmation of a name. Attention to detail was an important lesson for Banks because he was already building an herbarium, which grew so impressively that he had to renovate his home to accommodate it; like Sloane's, it eventually went to the British Museum. In 1768 Banks learned of plans for James Cook's first round-the-world voyage, and he had the wealth and political connections to insinuate himself into the expedition, which was co-sponsored by the Royal Society and the British Admiralty. After convincing Solander to go along, Banks financed a team of naturalists, even paying to refit a ship to accommodate them and their equipment, which he also supplied.[10]

There are great stories about the voyage and Banks's exploits, not all botanical, in places like Tahiti. He and Solander, along with two artists, a secretary, and an assistant, worked hard to record not only plant life but the animals, geography, and indigenous peoples they encountered. They amassed thousands of specimens, including many new species. The skilled botanical artist Sydney Parkinson illustrated over nine hundred plants, both finished works and pencil drawings with color notes. The routine they devised during the voyage was to collect all day, then return to the ship, where

Parkinson sketched the fresh material while Banks and Solander, with their secretary's help, wrote up observations and pressed specimens. There were times in Australia when collections were so copious the team could not dry them fast enough. Then they laid plants across the deck on sails during the day, changing the drying papers frequently. Careful attention was key to good preservation.[11] It was no wonder that when Banks faced a decision about Labillardière's herbarium, he respected the other collector's ownership: he knew what went into gathering thousands of specimens.

When Banks and Solander returned to London in 1771, Solander worked for Banks in describing the new species for publication. There was a great deal to do, with thousands of specimens to study, including not only pressed plants but a collection of fruits preserved in alcohol. There turned out to be more than a thousand new species, many that Solander found difficult to classify because they were so different from plants found elsewhere. Then there were the illustrations to prepare. Since Parkinson died before reaching England, others produced the final drawings and engravings. Publication kept being delayed, with some observers blaming Solander's laziness and interest in women and Banks's enlarging role in scientific administration. In any case, by the time Solander died suddenly in 1782, over eight hundred engravings were completed, but still nothing was printed. In fact, the massive work, in thirty-five volumes, was not published until the 1980s.[12]

Despite the difficulties, with Solander's assistance Banks became a competent botanist and continued to develop his herbarium, emulating Sloane in buying excellent collections like that of Philip Miller, head gardener at the Chelsea Physic Garden. Miller's herbarium was so large that it took two weeks to move. However, especially after Solander's death, more and more of Banks's time went into botanical management, fired by the Enlightenment idea of progress. He was informal adviser to King George III on botany, horticulture, and agriculture. Through Lord Bute's early influence, the king was avid about plants. Banks convinced George that the garden at his Kew Palace, which the king's mother Augusta had nurtured, should be developed into a research facility for growing plants with an eye toward turning them into economically important species.[13]

Banks never held a formal post at Kew, but he served for many years as unofficial director, with the gardener/botanist William Aiton in charge of day-to-day operations. Banks worked with a team of botanists led by Aiton on a catalogue of the plants at Kew; this grew over the years into a major reference work for botanists and is now maintained digitally. Banks also arranged for collectors worldwide to send specimens to Kew as well as seeds, seedlings, and cuttings to be grown there. Banks directed collectors in every part of the world, which explains why Kew has such a vast historical collection. The first plant collector Banks commissioned was Francis Masson, who collected in several countries, but his South African finds were particularly noteworthy.[14]

Humboldt and Bonpland

Unlike the voyages discussed so far, the one to Latin America undertaken by Alexander von Humboldt and Aimé Bonpland was a private undertaking—and a very fruitful one. When they completed the first portion of their exploration of South America, they had already amassed twelve thousand specimens from hundreds of species. Yet Humboldt recorded that they "were barely able to collect a tenth of the specimens met with." This might seem like hyperbole, but Bonpland also wrote of being almost overcome with the landscape's richness. Their remarks point to a problem every plant collector faces: what to select and what to ignore. In most cases, it is impossible to press material from every species; choices have to be made and weighed against the cost in paper, as well as in time for processing and description. Also, specimens are bulky, taking up space and weight in a collector's gear and later in shipping. Plants that are rare, unknown to the collector, or in an unexpected location tend to be selected. Moreover, humans, like many animals, are attracted to brightly colored flowers and foliage, and because of the human-size scale, they often miss tiny plants that would be obvious from the bird or insect perspective.[15]

Humboldt and Bonpland were experiencing biodiversity far greater than they were accustomed to in Europe's temperate climate, and their backgrounds made them well prepared to study the differences. Humboldt, who had a longtime interest in plants and geography, had recently inherited a sizable estate that allowed him to leave his job as a mine inspector in

Germany and move to Paris with the purpose of outfitting an expedition to the New World. Humboldt bought instruments to measure meteorological, astronomical, and geological phenomena and also equipped himself with materials for collecting plant, animal, and geologic specimens.[16] In Paris, Humboldt met Aimé Bonpland, a botanist trained at the Paris herbarium who was also ripe for adventure. When they could not attach themselves to a French government expedition, they went to Spain and convinced the king to issue them passports to his colonies in Latin America and the Philippines. This was a relatively easy sell since the explorers were only asking for access, not funds, and their plan fit with the king's vision of investigating his lands' botanical resources and how they could be effectively exploited.

After collecting in northern South America, Humboldt and Bonpland sailed to Cuba, where they explored around Havana and prepared their plants, animal skins, insects, fossils, and rocks for shipment back to Paris. They kept a small traveling herbarium with them to document what they had already collected; this was a relatively common and commonsense practice serving as a reminder of the species already acquired.[17] The pair returned to South America, to what is now Colombia, and made a treacherous trip into the Andes, finding the natural canal linking the Rio Negro and the Amazon. They continued to collect plants as they also measured temperature, altitude, air pressure, and other parameters.

Bonpland and Humboldt took careful notes and, in a move that was unusual for the time, numbered plant specimens both in a log and on specimen tags, also recording where each was collected, a description, and possible identification, as well as noting the altitude, a feature rarely considered earlier. Eventually they filled seven botanical notebooks. While they both collected and prepared specimens, Bonpland wrote most of the descriptions. In the first journal, 682 entries were his and only 9 Humboldt's. However, Humboldt was particularly taken with orchids and made drawings of them, presaging the orchid vogue later in the century. None of this work was easy. At one point they were so frustrated by losing specimens to insect damage and fungal rot that they made nature prints, going to the trouble of inking specimens and pressing them between sheets of paper (fig. 7.2). It was as close as they could sometimes come to preserving evidence of the plant itself.[18]

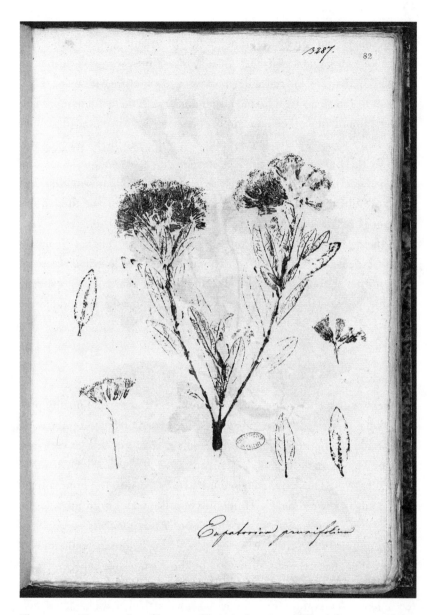

/3287. 82

Eupatorium prunifolium

Figure 7.2. Nature print of boneset (*Eupatorium prunifolium*) from a specimen collected by Alexander von Humboldt and Aimé Bonpland. (Library of the Institut de France. © RMN-Grand Palais/Art Resource, NY)

While in the Andes, Humboldt and Bonpland visited Bogota and stayed for two months with the botanist José Mutis, who was thrilled to share his herbarium, botanical illustrations, and expertise with them. They went on to Lima and then Mexico, spending a year there doing more collecting. Humboldt worked on mapping the areas he visited, learning about the country's geology and mining industry, and studying its ancient cultures. By this time, the pair had been traveling for almost five years and decided to head back to Europe rather than on to the Philippines. First they sailed to Cuba to pack up their specimens and equipment, and then took a ship north to Philadelphia, where they met many in the scientific circle centered there. They were invited to Washington, DC, by Thomas Jefferson and then spent a week with him at his home, Monticello. Jefferson was eager to learn what he could from Humboldt about Mexico since this was shortly after the Louisiana Purchase and the two countries now shared an extensive border.[19]

Reporting on the Expedition

On their return, Humboldt and Bonpland settled in Paris. The city provided what they needed for reporting on the expedition: good publishing opportunities, a flourishing intellectual milieu, excellent libraries and, perhaps most important, a first-rate herbarium at the National Museum of Natural History. They eventually published over two dozen volumes on the voyage, many coauthored by Humboldt and Bonpland, with most of the writing Humboldt's. The first publication, *Essay on the Geography of Plants,* outlined ideas that had come to him long before the expedition.[20] When young, he had been impressed by the relationship between plants and geography: different types of plants grew in different localities, and often similar plants grew in different places that had like characteristics, mountainous species being one example. Humboldt had seen this on his excursions in the Alps and on collecting trips with his early mentors.

Humboldt met the botanist and apothecary Carl Willdenow while still a teenager and learned botany from him at the time Willdenow had just published his flora of Berlin. At university, Humboldt became friendly with Georg Forster, who years before had traveled with his father on Captain James Cook's second voyage around the world. Like many educated elite of the time, Forster and Humboldt made a European tour, visiting the Netherlands, France, and England. This widened Humboldt's horizons, to say nothing of his knowledge of natural history. Later, during trips to mountainous areas as a mining inspector, he noted the influence of geography and soil type on flora.[21]

The Andes gave Humboldt a broader perspective because he could compare and contrast plants on different continents. Besides his mining experience and climbing in the Alps, he had already scaled Mount Teide, a volcano on Tenerife in the Canary Islands, and identified five plant zones: grasses near the peak, then shrubs, below them heath and pines, at still lower elevations forests of oaks, chestnuts, and other trees, and finally cultivated lands along the coast.[22] Factors determining these zones seemed to be elevation, soil, and water. After Humboldt descended from Mount Chimborazo in what is now Ecuador, his ideas were so clarified he immediately wrote a draft of the *Essay*. When he returned to Paris, he had the botanical artist Pierre Turpin draw a diagram illustrating how plant life varied with altitude in the Andes. The chart became a classic in the graphic arts, just as his ideas on biogeography are considered seminal in the field.

In 1824 Humboldt published a similar diagram in which he moved some plants to different elevations. Recent researchers, analyzing these images, have concluded that species names placed above the tree line represent "an intuitive construct based on unverified and therefore partly false field data that Humboldt constantly tried to revise in subsequent publications." The authors compared the plants in the diagram with specimens Humboldt and Bonpland collected. They found that Humboldt's material from above the tree line was mostly gathered at Mount Antisana, not Chimborazo, even though the diagram pictures the latter. The researchers visited the collection area and discovered that in two hundred years, the tree line shifted about 215–66 meters higher due to warming conditions. This reveals how old specimens can illuminate today's environmental issues, while also shedding light on how data were treated in the past.[23]

In Paris, Humboldt and Bonpland divided the expedition's specimens so each had as complete a set as possible. (As on many expeditions,
most species had been collected at least in duplicate.) Humboldt obtained
a pension for Bonpland from Emperor Napoleon to provide him with income as he studied the plants, and later Empress Josephine appointed Bonpland superintendent of gardens at her Malmaison estate. However, his
heart was in collecting, not writing about plants. After waiting years for results, Humboldt invited Willdenow to Paris to examine the specimens. His
former mentor worked on them for a few months, and then Humboldt pre-
vailed upon Willdenow's student Carl Kunth to move to Paris and take up
the project.

Eventually, Bonpland returned to South America with his herbarium.
He planned to bring the botanical field notebooks, but Kunth rushed to the
port at Le Havre and convinced him to cede them. Without this information it would have been nearly impossible to describe the plants adequately;
words as well as specimens were necessary. Keeping notes and specimens
together, and preserving both, continues to be an issue. They sometimes
become separated with the death or retirement of a botanist, and they are
often stored in separate facilities—herbarium and library—in some cases in
different institutions.

Kunth remained in Paris for seven years, eventually producing seven
volumes on the plants Humboldt and Bonpland had collected. He described over forty-five hundred species, about thirty-six hundred of them
new to science. One problem with this work was that while Humboldt was
still in the New World, he had sent specimens to Willdenow, allowing him
to describe anything new he found. After Willdenow died, two of his colleagues published some of his descriptions but would not share their work
with Kunth, whom they considered a traitor for leaving Germany and working in France, then a German enemy. Kunth described some of these same
plants as new species, leading to years of confusion—and revealing how
politics and botany intersect. Despite Kunth's work, not everything was
identified. In the twenty-first century, a new species in the potato family was
discovered in Peru, and the taxonomists describing it found that Humboldt
and Bonpland had collected it on their expedition.[24]

Eventually, Humboldt returned to Germany and toward the end of his life produced a five-volume opus, *Cosmos*, exploring the relationships among many forms of human knowledge. In these writings, Humboldt was influenced by the great German writer Johann Wolfgang von Goethe, whom he visited frequently when working in mines. He even took a three-month sabbatical to do experiments with Goethe, who had a deep interest in botany, kept an herbarium, and was intrigued by how plant forms related to one another. Both he and Humboldt were visual thinkers and espoused the Romantic idea of science and aesthetics as linked. Humboldt wrote about this in *Cosmos*, particularly in terms of mountain landscapes, but the same idea comes through in the care with which he had botanical illustrations produced. Most of the work was done by Turpin, and in all there were seven hundred plates in Kunth's seven volumes. Humboldt wanted the reader not only to learn about a plant but also to share in the experience of seeing its splendor.[25]

The beauty of plants has peeked out at several points in this chapter, from Mutis's illustrations through Banks's work at Kew to Empress Josephine's Malmaison. While power and wealth may have driven much botanical exploration, another important motivation was the search for beautiful and unusual plants for Europeans to grow in their gardens. Even from the days of Clusius, adorning gardens was important to many botanists, but this became a larger and more organized endeavor in the following centuries.

EIGHT

Gardens

A GARDEN IS A LANDSCAPE WHERE plants are purposefully grown, where a human is watching over them, planting and nurturing them. What plants are chosen depends upon the gardener's intention: a kitchen garden for food, a physic garden for medicinal plants, a pleasure garden for display, or one to work the soil, experiment with new plants, and create hybrids. Numerous studies indicate that gardening, and even spending time in a garden, can improve mental and physical health, in part through the pleasure of encounters with beautiful forms and fragrances. Some argue that the human attraction to flowers has a biological basis and is an adaptive advantage.[1] Just as animals with color vision are attracted to brightly colored flowers, so are humans, for whom the reward, although less tangible than pollen or nectar, focused instead on arousing pleasure, has similar roots in biology.

Because of the manifold ways plants can give pleasure, it is no wonder that gardens are deeply imbedded in human history. What is found growing in gardens is the result of millennia of plant exchanges around the world. Some gardens focus on native plants, but most have global collections. The pleasure in novelty and beauty has long fed the thirst for new plants to grow within the confines of a single garden, no matter its size. Tulips were unknown in Europe until the sixteenth century when they were brought from Turkey, just as sunflowers arrived at around the same time from the other direction, the Americas. Both were documented in early herbaria. Like hungry bees, gardeners were attracted to these plants' brightly colored blossoms and wanted more of them.

These two flowers became popular early in the age of exploration because they flourished in European climates, but plants from the tropics often did not. It was a matter not just of warmth but of soil type, watering
regimen, amount of light—all in the right combination. Patient trial and error won the day to the point that no part of the planet remained unchanged
by humans.[2] Gardeners have been responsible for a significant portion of
this, as have farmers. Both have been breeding and moving plants since pre-
historic times, leading to the spread of what we now term invasive species
and to habitat destruction, with herbaria providing evidence.

Gardening is about more than just finding new plants. Gardens, large
and small, are human-designed spaces. The disruption of the English Civil
War in the mid-seventeenth century led many wealthy Britons on the wrong
side of politics to spend those years on the continent, and the horticulturally minded used their time well. Robert Morison, later a professor at Oxford, was in charge of the king's garden in Blois, France. He learned about
the exotic plants displayed there and met the leading French botanists of
the day. Working for a wealthy and powerful garden patron revealed the
ever-expanding world of plants and the intricacies of tightly designed
French gardens. He returned to England with an herbarium preserving the
plants and knowledge he encountered in France. This contributed to his
work at Oxford University on plant description and classification.[3] Also in
exile at the time was John Evelyn, who visited botanical gardens in Paris,
Leiden, and Padua to learn about plants and toured private gardens to
broaden his understanding of garden design. This was a common practice
among elite gardeners, a way to learn from others and to allow the host gardener to display style, expertise, and access to the latest exotics.

On his return to England, Evelyn planned an ambitious project, an
encyclopedia of British gardening, *Elysium Britannicum*. By the late 1650s
he already had an outline for the work, but he never completed it. He
wanted to cover every aspect of garden design: what to plant, how to manage development, and garden maintenance. He even had a chapter on how
and why to create an herbarium as a reference for garden plants. Evelyn first
saw an herbarium while in Padua and was fascinated by how well preserved
the plants were, despite color loss. He was inspired to create his own, and

twenty years later he showed it to a friend, the famous diarist Samuel Pepys, who had never seen one before and was curious about this way of documenting plants.[4] Evelyn claimed that it was better than any printed herbal. Obviously, there were many people engaged with plants—explorers, botanists, gardeners—who felt this way, and as the years went on, collections continued to be made and were more likely to be preserved as appreciation for them grew.

Women Collectors

A gardener who carefully documented her collection was the Duchess of Beaufort, Mary Somerset, a good friend of Hans Sloane. Somerset combined expertise in gardening, horticulture, and botany.[5] Her herbarium included varieties of common garden plants and rare exotics, often tropical plants, which did not usually do well in Britain. The solution was a "stove," a greenhouse with a heating system. Somerset was one of the first in Britain with one and had a reputation for coaxing difficult species not only to flourish but to bloom.

Somerset lived at a complicated time in British history, when kings were deposed and even executed. Since she came from an influential family and married into two others, she faced many problems beyond the confines of her gardens. In midlife she suffered from what was then called melancholy, finding her way out of it through correspondence with Joseph Glanvill, a philosopher and clergyman who was a forerunner of today's interest in the wellness benefits of gardening. He argued that learning about the natural world could help to overcome depression. Somerset, encouraged by this idea, put it into practice by spending more time gardening, observing plants' growth conditions closely, and studying botanical references. It was at this point that she began to seriously grow exotics and note their characteristics.[6]

Having ties with a number of collectors, Somerset received plants from India, China, Africa, and the Americas. She obtained the freshest possible materials to nurture because she had the means to spend lavishly on her garden. One massive shipment from Barbados with seeds, cuttings, saplings, and several larger trees had to be divided among five ships.[7] Somerset was also interested in native plants and financed a collector to go to the mountains of northwestern Wales, but it was the exotics that made her important to other botanists and collectors.

Somerset developed her understanding of botany through a network that included Hans Sloane, James Petiver, William Sherard, and John Ray, who used her herbarium in his research. She had access to important botanical references; in her notes she referred to not only Morison and Ray but even to van Rheede's work on Indian plants. Through Petiver she was linked to his collectors, and he enjoyed visiting her gardens because of the wonders he saw growing there. Somerset was assisted in her work by William Sherard, a botanist who had worked at Oxford and whom she hired as her grandson's tutor. Schooling her in botany, he used his connections to add many exotics to her garden and developed her herbarium as she worked alongside him.[8]

The duchess kept track of the many plants she cultivated, including familiar ones. In her herbarium there are pages of anemone flowers, varieties that have long since disappeared and for which the collection is a permanent record of their existence (fig. 8.1). It is not surprising that this twelve-volume herbarium is now part of Hans Sloane's collection. Later in life, when Somerset moved from her estate in Beaufort to a house and garden in Chelsea, she was his neighbor. Concerned to get plant names right, she corresponded with Sloane and others, admitting that neither she nor her gardener knew Latin, yet Sloane thought so highly of her cultivation skills and facilities that he had her grow medicinal plants for the Royal College of Physicians.[9] She traded plants and gardening information with Sloane, with gardeners at the nearby Chelsea Physic Garden, and with Jacob Bobart the Younger at Oxford.

Somerset documented her successes not only in her herbarium but by having her plants drawn by artists, including Everhard Kick, who had painted the Jamaican plants in Sloane's collection. In one case, he depicted a plant from the Cape of Good Hope in South Africa before it was introduced into British horticulture, suggesting that Somerset received it directly from the collector, a sign of her status in the botanical network. She

Figure 8.1. Anemones from Mary Somerset's herbarium, each enclosed in its own sheet of paper. (© The Trustees of the Natural History Museum, London)

used Kick's paintings and those of others as templates for embroidery. She was a skilled needle worker, as were many upper-class women of her time, and flowers were a favorite subject. Since women were limited in their educational opportunities, they used such outlets to grow intellectually through what were deemed feminine arts. They considered this another form of study, a way to learn about plant form and structure, an adjunct to working in the garden or creating an herbarium. Floral embroidery became a continuing theme in women's botanical interests, and there were botanical art books that doubled as pattern books. Mary, Queen of Scots, some of whose floral embroideries still exist, used patterns based on illustrations in Mattioli's herbal.[10]

Years later, although the constraints remained, the number of women horticulturists had grown. The Duchess of Portland, Margaret Bentinck, was another wealthy woman who used plants as a way to develop her intellect, her aesthetic sense, and her gardens. Bentinck collected on a grand scale, everything from plants to minerals, shells, ceramics, and paintings. These were auctioned soon after her death and little remains of them. Like Somerset, she hired a leading botanical artist, in this case Georg Ehret, to document her plants in watercolors and also to teach her daughters painting. Bentinck was a patron to the philosopher Jean-Jacques Rousseau, who later in life studied botany, believing it calmed the emotions by focusing the mind on something outside itself. Botany became an important manifestation of his interest in nature, usually in common species rather than exotics and horticultural "monstrosities."[11]

When Rousseau spent time in England, he visited Bentinck's estate and botanized with her. As he did for several of his patrons, he gave her a portable herbarium; he considered a plant collection a way to reinforce botanical knowledge (fig. 8.2). A Swiss botanist had taught him how to press plants, and he later wrote about the process for Madeleine Delessert, wife of a French financier. She sought Rousseau's advice about teaching her daughter botany. In response, he sent her eight letters on the art of observation and how to compare plant forms, ending with a section on creating an herbarium. He also made an herbarium for Delessert. There is little information on how Rousseau's work influenced her daughter, but her son Benjamin, who

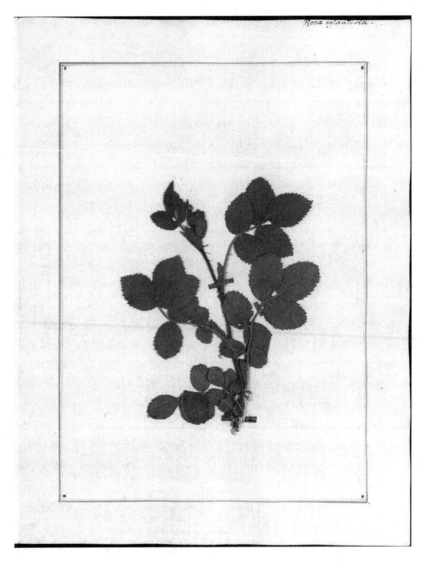

Rosa eglanteria.

Figure 8.2. Specimen of sweetbriar (*Rosa eglanteria*) from an herbarium made by Jean-Jacques Rousseau for Mademoiselle Julie Boy-de-la-Tour, a relative of Madame Delessert, for whom he had also created an herbarium. (Zurich Central Library)

became a leading banker, was fascinated by botany and developed a large herbarium and botanical library cared for by curators in the tradition of Hans Sloane.[12]

As a member of the nobility, Bentinck had a wide social circle beyond intellectuals like Rousseau, but her closest confidante was Mary Delany, the wife of a Dublin clergyman. Although Delany participated in the London social scene at times, she lived a good life in Ireland, creating a garden and designing elaborate needlework patterns with highly realistic representations of flowers. When presented at court, she wore a gown with two hundred embroidered flowers, all identifiable by species; she had obviously learned plant form as well as art. After Delany was widowed, she spent months each year staying with Bentinck, also by that time a widow. They studied Linnaean botany with the Reverend John Lightfoot, who organized Bentinck's specimens, collected for her, and served as her chaplain.[13] To occupy herself after a foot injury when she was seventy, Delany painted bits of paper and cut them into forms to create flower collages. These mimicked botanical illustrations in that a single stem in flower or fruit was placed against a plain background, though painted black. In ten years Delany completed 985 botanical cutouts, labeled with the Linnaean name for each species. When King George III and his wife Queen Charlotte visited Bentinck and Delany, they marveled at the cutouts and Delany soon had access to plants from the garden at Kew.

Like her embroidery, Delany's cutouts are notable examples of how women combined botanical knowledge with the arts, and of how herbaria influenced that art. The collages are similar to herbarium specimens in that they have more depth and texture than an illustration. There are even cases where Delany added real leaves to a work. Through her studies with Lightfoot, Delany was familiar with specimen preparation and so with arranging a plant on paper, flattening it out, and making sure its essential features were displayed. She did something similar with her cutouts and often depicted both sides of a plant's leaves, common practice in mounting specimens. Botany and art drove her to look closely and become more aware of flower form and specimen preparation.

Nurserymen

As the examples of Somerset and Bentinck illustrate, large sums of money were lavished on gardens. Over the centuries, as the demand for plants grew, gardening became a business. The wealthy employed managers for their estates and these men in turn hired those who actually did the work. A new form of business evolved: the nursery, where gardeners could buy equipment and novel plants. Nurserymen often dealt directly with collectors or with plant importers, many based in the Netherlands, with its broad shipping connections. There was also transmission of information and plants, especially seeds, across Spain, France, the Netherlands, and Britain, with each country's plantsmen having different sources for exotics. Plants from North America and South Africa became particularly popular because their native ranges had climates more similar to Europe's than those from India or South America.

On the outskirts of London nurserymen cultivated plants in large numbers and also sold the seeds they harvested. They experimented with better ways of nurturing plants, and Thomas Fairchild was the first to have a blooming horse chestnut tree in Britain, almost two hundred years after Ogier Ghiselin de Busbecq brought its seeds and specimens from the Middle East for Clusius and Mattioli. Fairchild was also the first to create a hybrid between two plant species. There is an herbarium specimen at Oxford of "Fairchild's mule," a cross between sweet William and carnation pink—an example of how herbaria can document otherwise fleeting botanical accomplishments.[14]

As with any business, nurseries needed ways to advertise their wares. Some used an herbarium as a sales catalogue; it was portable and provided proof that the species was on hand. The tactic appealed to sophisticated gardeners who also documented their plants this way. Other sellers simply printed up lists of what they had in stock, perhaps with a brief description of each. A few added woodcuts picturing their wares, and for the most elite clients, there were colored engravings, as in Robert Furber's *Twelve Months of Flowers,* adorned with plates of flowering plants he sold, organized by the month they bloomed.[15]

Sometimes the advertising was more subtle. Mark Catesby, perhaps the best-known British naturalist who visited colonial North America,

made two trips there. On the second, begun in 1722, he gathered a large collection of plant and animal specimens, as well as notes and drawings he
used in his two-volume *The Natural History of Carolina, Florida and the
Bahama Islands.* He produced its etched illustrations himself because he
could not afford to hire a professional. To make ends meet, he was involved
in London's nursery trade, where North American plants were popular.
A number of his descriptions include the names of nurserymen who successfully grew the species. Since only the wealthy could afford these books,
they served almost as catalogues for American exotics aimed at a receptive
clientele.[16]

John Bartram

Trees and shrubs were much desired in horticulture to add grace to the landscapes of large estates, where the British gardening style included what appeared to be natural settings; also, their longevity made them worth the
expense. North American colonies were fertile hunting grounds for novel
trees. Peter Collinson was an avid gardener with botanical knowledge and a
large network including, as mentioned earlier, Linnaeus, Clifford, Bentinck,
Sloane, and many aristocratic gardeners. Since Collinson was in the textile
business, he already had mercantile contacts in the American colonies, especially among fellow Quakers. One of them put him in touch with John Bartram, a Philadelphia farmer who had cultivated and experimented with
medicinal herbs and was willing to engage in trade. Since Bartram's botanical knowledge was limited, Collinson devised a system in which Bartram
made at least two specimens of each species, one for himself and the other to
send to Collinson along with seeds and seedlings. They employed a nu-
merical code so Bartram could label his specimens when he received Collinson's list of names. If the latter could not identify a plant, he sent it to other
botanists for assistance, including Johann Gronovius in the Netherlands and
even Carl Linnaeus, so these men became familiar with Bartram's work.[17]

   To meet the demands of customers, Collinson sent Bartram paper for
pressing specimens, a microscope, and books on botany. Bartram had already been introduced to botanical Latin by James Logan, a Philadelphia

merchant and political figure with an interest in botany. The year after Linnaeus's *Systema Naturae* was published in 1735, Logan had a copy and later taught Bartram the system. By the time Pehr Kalm visited him in 1749, Bartram had a command of North American plants and was traveling broadly in the colonies at Collinson's behest seeking new species to send to England. Bartram eventually explored from northern New York into western Pennsylvania and as far south as Florida. Bartram's specimens include many from his travels, though the notations are not very specific, such as "This I gathered on the Katskill Mountains."[18]

Over the thirty-five years Bartram and Collinson corresponded, there were fifty-seven gardeners who sponsored the boxes of seeds and seedlings that Bartram sent to Collinson for distribution. The first and most passionate was Lord Robert Petre, a young man obsessed with improving his estate with plants, particularly trees, from around the world. The first shipment from Bartram included three thousand black walnuts, over a peck of dogwood berries, two pecks of red cedar berries, and thirty-two hundred pin oak acorns.[19] Some of Bartram's specimens are now in the Sutro Library in San Francisco, among the plants in Lord Petre's sixteen-volume herbarium. There are two volumes with Bartram's plants, including a number with brown paper labels in his handwriting. Many specimens have long, pre-Linnaean names in beautiful script as well as references to works by Ray, Tournefort, and others; it was a well-curated collection. Historians attribute these annotations to one of the horticultural advisers Petre employed. In 1742, Petre died at twenty-nine. His heirs retained his library and herbarium but sold his rare living plants, with many elite British gardeners vying for them, including Collinson, of course.

The herbarium remained in the family until the late nineteenth century, when Adolph Sutro, a former San Francisco mayor, bought it during a European tour aimed at creating a world-class book collection for his city. That is how an eighteenth-century Philadelphia farmer's plants ended up in San Francisco after making two trips across the Atlantic. In the late twentieth century, Philadelphia botanists studied the Bartram portion of the Petre herbarium, identified the specimens, and labeled them with up-to-date names. They also wrote of the collection's historical significance, with its early observations on introduced species, including henbit and common starwort from Europe and green carpetweed from the American tropics.[20]

Experimentation and Acclimatization

Growing plants in gardens, botanic or personal, could be a challenge, and "acclimatization" became an important goal: to have a plant from one climate or ecosystem thrive in a different one. Mary Somerset did this on a small scale, but to introduce a plant into agriculture or horticulture required much more time and a larger operation. In 1652, the Dutch created a botanical garden at Cape Town, a trading post for the Dutch East India Company. Like many colonial gardens, its function changed over time as the needs of the colonial government did. At first, it produced fresh vegetables, mostly European varieties, to supply company ships stopping at the Cape. Over time, many European and Indian food plants were grown. Later, the garden was enlarged, with a dedicated space for flowering plants and eventually a program for acclimatizing promising species collected from areas of Asia involved in trade with the Dutch.

In the 1690s, Henrik Bernard Oldenland explored parts of South Africa and collected plants for the Dutch trading company. Later he was superintendent of its Cape Town garden and created a fourteen-volume herbarium from the plants he collected, especially those he then cultivated in the garden. The specimens serve as a record of these plants and were sent back to the Netherlands and acquired by Johannes Burman after Oldenland died. Burman's son later took them to Sweden so Linnaeus could study them and describe species based on them, making the collection historically important. Benjamin Delessert, son of Rousseau's friend, eventually bought the herbarium and willed it to the Geneva herbarium after its travels from Africa to the Netherlands and Sweden, then to France and Switzerland—each leg adding to its historical and scientific richness.[21]

Successful cultivation at gardens like Oldenland's made more plants available for distribution to other colonies and to the home country. There were studies in the gardens on whether or not related species, or ones that were thought to be related, had similar properties. This issue arose with a cinnamon-like tree in Peru that was ultimately a disappointment to those looking for a new source of the expensive Asian spice. It took two centuries to identify them as different species, since the two grew on different sides of the globe, and specimens that reached Europe weren't examined side by side. A French botanist who had studied Asian cinnamon trees and then traveled to Peru was the one who finally identified them as different species, having seen live plants of both. Ginseng from Asia, which has a root resembling a human form, had been highly prized in Europe for centuries. A similar plant was found in North America, spurring hope that this supply source would provide ginseng at a greatly reduced price. But did it have the same medicinal properties? For the work of determining this, there was no substitute for growing the plants and testing their potency, with specimens needed to document precisely what was grown. Eventually, the two plants were deemed different species, though with similar medicinal properties.[22]

The story of *Cinchona lanceifolia,* source of quinine and native to the Andes, illustrates how difficult it was to work out a plant's chemistry. The Spanish struggled to learn both to extract the active ingredient from the tree's bark and to find cinchona varieties that were particularly rich in it. In 1783, when Spain was seeking to better use its colonies' botanical resources, the viceroy of New Granada asked one of his governors to supply "skeletons," herbarium specimens, of the local cinchona trees, from which the precious bark was harvested. The governor was suspicious of this request, judging rightly that the botanists in Bogotá, the capital, wanted to compare these specimens with ones from trees they were growing. There was a great deal of debate about the quinine level in different trees, and even how to measure it, with some arguing for chemical tests and others seeing medical effectiveness as the only valid measure. One problem was that there were several species of the *Cinchona* genus, all with varying quinine levels; purification also resulted in fluctuating yields.[23]

While South America was the source for cinchona worldwide, other countries were attempting to raise the trees in their colonies. The Royal Botanic Gardens, Kew, grew cinchona seeds spirited out of Ecuador by Richard Spruce, a nineteenth-century British collector. The plants flourished at Kew and were sent to Britain's botanical gardens in the tropics.

Kew was also offered seed from a British expatriate living in Peru and turned him down, not knowing the origin of the seeds, but the Dutch were willing to take a chance and grew them at their botanical garden in Java. These seeds were identified as from a different species, now named *Cinchona calisaya,* that produced more quinine, allowing the Dutch in Java to corner the market until World War II. While Britain never competed successfully in the international quinine trade, the Kew experiments led to the transfer of *Cinchona succirubra,* or red cinchona, to Britain's colonies in Africa and Asia that were plagued by malaria, so they could avoid paying the high Dutch prices.[24]

In the nineteenth century, Kew was the center of a network of British colonial gardens, including ones in Calcutta, St. Vincent in the West Indies, and Sri Lanka. The Dutch and French also had such satellites. While the gardens served as pleasant retreats for a select few, these sites were designed to develop native species into agricultural products, test exotics for the same purpose, and acclimatize plants that could then be sent to the mother country or to other colonies. The gardens usually housed herbaria for sorting and storing specimens and, most important, sending home specimens of everything collected. Numerous books on native species were written based on collections such as Nathaniel Wallich's rare plants of East India. Director of the Calcutta garden for thirty years, he was particularly prolific, going on numerous collecting trips and overseeing other collectors. On a survey trip to Burma, he amassed sixteen thousand wood specimens and found five hundred new tree species.[25]

Britain had long ago decimated its own forests, while timber need only increased. Trees became an important element of imperial exploitation in India and Southeast Asia. In India forest destruction was so intense it caused fears of future shortages and resulted in environmental changes: cutting down trees led to changes in local weather patterns, loss of soil fertility, and consequently environmental deterioration in large areas of the country. Indian botanic gardens were used as experimental stations to study trees that might be grown on plantations to produce lumber for the future and for reforestation projects.[26] At one point the Singapore Botanic Gardens focused on growing large numbers of plants, particularly Brazilian rubber trees, to be distributed throughout the country and to other parts of the British Empire, while Asian teak was grown on plantations in Africa.

Many species, such as rubber, were successfully moved from country to country by the British and other colonial powers, as mentioned earlier. The French brought nutmeg and clove trees from the Molucca Islands to Madagascar and Zanzibar, successfully cultivating them there. Through the slave trade maize, cassava, and sweet potato were brought across the Atlantic to West Africa. The Dutch grew Latin American cacao in Indonesia and Sri Lanka. The British transplanted nutmeg to Sri Lanka, Malaysia, and even Grenada, where most is now produced. Vanilla, native to Mexico, was cultivated in Madagascar in the nineteenth century, and the island remains a major supplier. For this extractive colonialism indigenous peoples paid a high price in terms of forced labor on plantations and destruction of entire ecosystems, and these problems became more intense as additional species were cultivated when colonization grew more intense in the nineteenth century.[27]

Managing Exploration and Collecting

WHILE EXPLORATIONS AROUND THE world had been going on
for three centuries, the pace and scope intensified in the nineteenth century,
fueled by economics and politics as well as science. There were still many
uninvestigated areas. Even well into the nineteenth century, sub-Saharan
Africa remained relatively unknown botanically, apart from South Africa,
which attracted attention because of its species diversity and its position on
trade routes to the Far East. Its European-like climate made the flora attractive to European plant hunters and held less of a malaria threat. The Dutch
were particularly interested in regions having many geophytes, plants with
underground structures such as bulbs.

It was only later in the nineteenth century, when the British and
Dutch had grown adept at cultivating cinchona and extracting quinine from
its bark, that exploration and colonization became practical in Central
Africa. There had been some earlier forays. In 1748, the French botanist
Michel Adanson began five years in Senegal, a West African area then under
French influence as a site for slave trading. He brought back pressed
plants, seeds from a thousand species, and wood, resin, and gum samples.
Most notably, he described the baobab tree with its nutritious fruit and
remarkably broad trunk that served as a water reservoir. But before he
had an opportunity to formally publish a name, Linnaeus christened it
Adansonia, based on information he received from one of Adanson's
colleagues. The honoree was annoyed—he would have been happier to be
perpetuated in botany for naming it. He did write two books on African

plants, using a natural classification system influenced by both Tournefort and Linnaeus.[1]

Off the coast of East Africa, Mauritius was for many years a French stronghold, and its acclimatization garden was the oldest tropical botanical garden to provide novel species for French colonies worldwide. From the mid-nineteenth century, Belgium controlled a large area in Central Africa, and the king commissioned explorers to find resources to exploit, including plants that were sent back to the herbarium at the Meise Botanic Garden. It is estimated that 85 percent of the collections from this area are in Meise, now the center of a project to produce a flora, a catalogue, of Central African species.[2]

David Livingstone's name is well known as an African missionary and explorer, but the botanical aspect of his story is usually underplayed, with the emphasis on wild beasts and dangerous terrain. The most notable collections were made on the Zambezi Expedition of 1858–64 when a party of Europeans and Africans journeyed up the Zambezi River. John Kirk was the British botanist and physician who did most of the collecting, contributing to the gradual growth of knowledge about African botany. In 2012 the Royal Botanic Gardens, Kew, completed publication of a flora of tropical East Africa, including the area where Kirk collected, a part of Africa with several former British colonies, so it is not surprising that the British took the lead on this project, which began in the 1930s. Plans had to be delayed after World War II because of the Berlin-Dahlem Herbarium's destruction. It held many African specimens that had to be re-collected. While Germany was never a major colonial power, it had many international commercial interests and its botanists were active in acquiring specimens. Many were experts on African flora, financing collectors and acquiring specimens through trade with French and British botanists of like mind and readier access.[3] There are many ways to build expertise, and the world of herbaria is full of examples of exchange.

On the other side of the globe, the German collector Johann Preiss explored Western Australia, sometimes in collaboration with colonists who made a living sending plants back to Europe. James Drummond immigrated to Western Australia in 1829 when the area was a frontier region and conditions were difficult. Plant collecting became an important source of income for him, despite the obstacles. Not only was it challenging to survey a terrain with thorny, often impenetrable, flora, but even something as simple as paper was often impossible to procure. Supplies usually came from Britain via Cape Town to Western Australia, and plants had to be shipped back the same way. Sailings were infrequent, and if a shipment of seeds did not make it to port in time, an entire season's work might deteriorate before the next sailing.[4]

Other parts of Australia were investigated earlier, particularly along the coast. French explorers like Jacques Labillardière made early collections, but the British were determined to take the lead. In 1787 they founded a colony in Sydney, and in 1801 sent an expedition to circumnavigate Australia and map its coast. In his role as a botanical organizer, Joseph Banks chose Robert Brown as botanist and Ferdinand Bauer as artist for the voyage. They made a good team, even working together on their return to describe and illustrate their collections. Unfortunately, Brown was slow at writing and the British government was unwilling to finance the publication of many illustrations. Just fifty-seven plates made from Bauer's over fifteen hundred drawings were published at the time, and Brown managed to produce only a shortened version of his work.[5] However, the real riches of the pair's explorations are the specimens, most of which are, like so many colonial collections, not in their place of origin but in the mother country.

While Brown's major herbarium went to the British Museum, a substantial number of duplicates was deposited at the Royal Botanic Garden, Edinburgh. Its herbarium was also receiving large collections from physicians who had studied botany in Edinburgh and were encouraged to continue their interest in plants. Some alumni worked for the East India Company, often as physicians, and became responsible for botanic gardens the company was developing in India and East Asia to further the British colonial enterprise. These men collaborated with indigenous doctors and apothecaries to learn about local plant uses and studied species long-term in the gardens they planted. They also collected and sent specimens not only to company headquarters in London but also to Edinburgh. This became more common over time, as did accompanying specimens with watercolor illustrations of the plants done by local artists who adapted indigenous styles to this genre (fig. 9.1). Large collections at Edinburgh and Kew suggest that the watercolors became very popular. They were used for reference by botanists in part because it was difficult to prepare specimens in such a hot, humid climate.[6]

Imperial Oversight

One issue that arose often with plant collections, particularly in the nineteenth century, was who had sufficient knowledge to describe species. Bruno Latour's idea of centers of calculation is relevant here. As specimens were sent back to Europe and became concentrated there, European botanists saw themselves as in the best position to describe new plants; they had vast herbaria as reference material. Since Joseph Banks built Kew into an important center of British botany, it was natural that those at Kew would take the lead. William Jackson Hooker arrived there as its first full-time director in 1841. He brought with him a large herbarium that he left to his son, Joseph Dalton Hooker, who was named his successor and sold his father's herbarium to Kew, a way to augment his meager Kew salary. He then continued Kew's work in amassing more specimens.[7]

The Hookers received material from collectors throughout the British Empire, with Australia and New Zealand as particularly attractive hunting grounds, being newly settled and with a flora so different from any other. Father and son made it clear that those who gathered specimens were not taxonomically sophisticated enough to describe new species. Some collectors were fine with this, but others bristled as they developed botanical expertise. The so-called closet botanists holed up in their offices kept writing to dissuade those with such audacity. One argument was that field collectors tended to name too many species. Without sufficient specimens for comparison, they saw a plant with different traits as a novelty, whereas a botanist with a large collection at his disposal could appreciate that the same plant was just a variety of a species that the collector never saw. Professional taxonomists sought to control the number of new species named in part so they would not be overwhelmed.[8]

Figure 9.1. Illustration by an unnamed artist of royal poinciana (*Delonix regia*) among a collection of watercolors created at the Garden of Dapuri in Mumbai, India, in the mid-nineteenth century. The tree is native to Madagascar, but by this time had been planted in many tropical areas as a street tree. (Archives of the Royal Botanic Garden, Edinburgh)

In Melbourne, Australia, the German botanist Ferdinand von Mueller had developed a good botanic garden and herbarium and wanted to write the flora of Australia himself, but he refused to travel to England to consult with the Hookers and their colleague George Bentham at Kew. They were describing Australian plants and deemed Kew's collection essential as reference material when the project began in 1858. Mueller finally ceded control to the Kew botanists and was listed as coauthor. A good deal of Melbourne's collection was packed up bit by bit and sent to Bentham for study and then systematically returned. This might be the largest herbarium loan ever.[9] Eventually the tension over classification subsided as botanists like Hooker kept getting sent the same plants again and again; they left it to those on-site to sort things out.

It was not just botanists in botanic gardens who dealt with collectors. The Royal Horticultural Society (RHS) sent out plant hunters with a different goal: not necessarily to discover new species but to find ones that could be cultivated. Robert Fortune had been a gardener and the first recipient of a certificate in horticulture from the RHS, a sign of the professionalization of gardening. He was sent to China in 1843 to find species with potential as garden plants, particularly trees and shrubs. This meant he did little exploration of uncharted territory, instead visiting nurseries and gardens in the few areas open to foreigners. He sent pressed plants to John Lindley, the RHS secretary, but his major work was collecting enough seeds to grow each species in sufficient numbers to evaluate their properties. In some cases, if the seeds were difficult to collect or to germinate, seedlings were sent.

Transporting live plants on a long ocean voyage was a major undertaking that was made easier by the invention of the Wardian case in the late 1830s. It was essentially a box with a glass lid that sealed in moisture, let in light, and protected plants from desiccation and the salt air onboard ships. The case was not perfect, and some shippers were more successful than others, but generally a higher percentage of plants survived their travels in these containers. On his second expedition to China, Fortune was working for the East India Company, which aimed to grow tea in India. He made use of Wardian cases and Chinese expertise to bring tea plants to India. While he is often described as having "stolen" tea from the Chinese, there is a difference of opinion about this, as with many examples of plants moved from country to country.[10] Fortune was successful because he recruited Chinese tea growers, who accompanied his shipments to India and directed cultivation there. In addition, Fortune's careful observations confirmed that green and black teas were derived from the same plant, the difference resulting from how leaves were processed.

Another plant explorer in China was Armand David, a French priest sent to Beijing as one in a long line of missionaries who also botanized. Because he already had expertise in plant collecting in Europe, French botanists enlisted him to collect in China. Like Fortune, David studied the Chinese language so he could communicate with those he met and, like Fortune, he adopted Chinese dress. These men realized they could do little without aid from locals and wanted to fit in as much as possible. Over the course of several years, David made three major expeditions, collecting thousands of specimens he sent back to Paris. Besides plants, he gathered animal skins, including those of pandas and a deer species that was even then near extinction but kept in imperial preserves: what became known as Père David's deer. As for plants, David's most notable find might be the handkerchief tree, a dogwood-like species with large white leaf structures that hang down from the flowers, making the tree look as if it's decorated with small white cloths.[11]

While David discovered the tree, it took two other plant collectors to bring it into cultivation. Since David and his sponsors were more interested in identifying new species than in cultivating them, specimens were more important than seeds and seedlings. The same was true for the British collector Augustine Henry, who worked as a surgeon and plant collector for a shipping company in several parts of China. These included mountainous areas of interest to Europeans because of the temperate climate and rich plant diversity. During his almost twenty years in China, Henry collected over 150,000 specimens. Uncharacteristically for the time, he made it a practice to name the indigenous collectors who worked with him. He encountered David's tree but collected only specimens, not seeds. Ernest Wilson, a professional plant collector working for a nurseryman, visited Henry specifically to find out where to locate *Davidia* and gather its seeds. Henry

gave him a crude map that did lead Wilson to the tree, but it had been destroyed. Eventually he found *Davidia* at another location and collected
large quantities of seeds, finally making cultivation possible.[12]

American Exploration

Like Australia, the western United States was largely unexplored at the beginning of the nineteenth century. In 1804 Meriwether Lewis and William
Clark set out on their expedition into the recently acquired Louisiana Purchase territory at the direction of President Thomas Jefferson. Their aims
were to map areas they visited, evaluate resources, learn more about the indigenous people living there, and cross the Continental Divide. They made
significant accomplishments and, as was the case with many government-
sponsored expeditions that followed, they collected plant and animal specimens. Jefferson considered this so important that before they set out he
had Lewis study botany and specimen preparation with Benjamin Barton,
author of the first U.S. botany textbook.[13]

Barton also received the specimens brought back from the expedition
and hired Frederick Pursh, a German botanist working in Philadelphia, to
study them. Before finishing the project, Pursh went to England, taking a
portion of the specimens as well as his drawings and notes. He eventually
published descriptions of many new species from the Lewis and Clark material as well as from other collectors. Though it was important to have
these species described, Pursh developed a dubious reputation for usurp-
ing and describing plants collected by others, including those of two British
botanists. They traveled in the western United States and sent their collections back to England, where Pursh was able to see and describe some of
their material before they could. This botanical sin is not unique to Pursh—
the lure of naming new species is a strong one.[14]

Most of the Pursh haul of Lewis and Clark specimens was eventually
bought at a British auction by an American botanist, who donated them to
the Academy of Natural Sciences in Philadelphia (see fig. I.1). The specimens Pursh didn't take were stored at the nearby American Philosophical
Society and forgotten until Thomas Meehan, the academy's botanical cura-

tor, rediscovered them near the end of the nineteenth century and organized the specimens. Such resurrections are relatively common with old collections, which are looked at with different eyes over time. The specimens are now at the academy, though they are still officially owned by the society, a type of arrangement not infrequent with old collections.[15]

A number of government-sponsored expeditions after Lewis and Clark's were directed by the military and were designed not only to explore what had become a vast nation about which little was known, but also to assert political and naval power. Among the most significant was the United States Exploring Expedition of 1838–42, which circumnavigated the globe, visited Antarctica, and included an overland expedition along the West Coast from the mouth of the Columbia River down to San Francisco. Often called the Wilkes Expedition after Charles Wilkes, who led it, this was a much larger undertaking than Lewis and Clark's and was essentially a naval operation, equipped with a team of scientists and artists to record the natural history of the areas visited. Along with zoological, geological, and anthropological materials, over fifty thousand plant specimens as well as live plants in Wardian cases arrived in Washington, DC: almost twenty tons of shipments. They presented a massive problem in how to process and report on them.[16]

The specimens ultimately ended up in the newly founded Smithsonian Institution, while Wilkes directed the documentation effort, distributing materials to experts in various fields. For the botanical work, he turned to John Torrey of New York, who had already described new species from earlier expeditions, and his protégé Asa Gray, recently appointed professor of natural history at Harvard University. Torrey was willing to study the specimens from the West Coast. Gray agreed to take on the rest of the collection, if he could first travel to Europe and examine plants in the great herbaria there. Wilkes balked: American plants had to be studied in America.[17]

Holding a viewpoint similar to the Hookers', Gray persisted, explaining that the United States had no large herbaria with sufficient specimens from around the world to allow him to make sense of the new collections, comparison being at the heart of classification. Only in places like Kew in

London and the National Museum of Natural History in Paris would there be a sufficient range of plants, many collected on European expeditions over three centuries. Wilkes finally relented, and for the second time in his career, Gray went to Europe. On an earlier trip, he had been party to what would now be considered an herbarium felony. The botanist John Lindley arranged for Gray to view a collection made in the Carolinas in the 1780s and attributed to the plantation owner Thomas Walter. Gray was even allowed to snip off small portions of the specimens for future reference. There is no evidence of similar pillage on his second trip, though there is a record of Robert Brown removing a portion of a William Dampier specimen during his work on Australian plants (fig. 9.2).[18]

Despite the massive Wilkes haul facing them, Torrey and Gray were still soliciting and receiving specimens from other collectors. There was a race to describe new species from the western United States involving European botanists like William Hooker and George Bentham, who were studying Canadian plants and also had collectors working in the western United States. Torrey and Gray saw it as a patriotic duty, as well as a matter of botanical pride, to describe as many species as they could. A case in point involves the intricacies of naming plants and a third botanist, William Darlington of West Chester, Pennsylvania. Like Torrey and Gray, he trained as a physician, but unlike them he practiced medicine throughout his life with botany as his avocation, though a serious one. His advice was sought on the Wilkes expedition, and he corresponded frequently with prominent botanists including Torrey, Gray, and William Hooker. Darlington created his own herbarium and wrote a flora describing plants from his home county that became a classic.[19]

In the 1830s, in gratitude for specimens sent by Darlington, the Swiss botanist Augustin Pyramus de Candolle named a genus related to mimosa after him, *Darlingtonia.* However, George Bentham later assigned the plant to another genus, so Darlington "lost" his genus, a common occurrence in botany as taxonomists reexamine the work of others in the light of new evidence. In the 1840s, Torrey named a different genus after Darlington, but he did it based on a rather ragged specimen, indicating the impatience to publish descriptions of new species from the West. When one of Torrey's collectors arrived at his home with a bundle of California plants, Torrey found a better

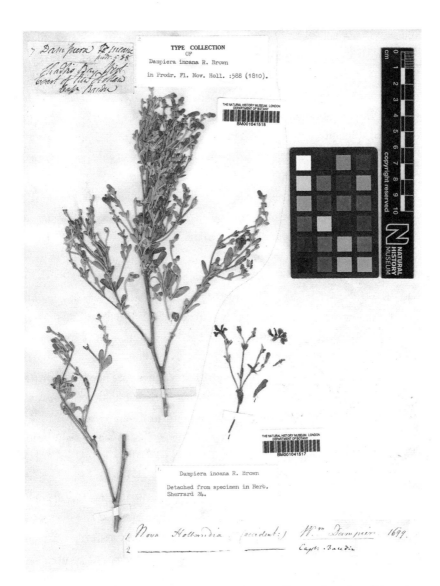

Figure 9.2. Herbarium sheet with two specimens of hoary dampiera (*Dampiera incana*) collected by Nicolas Baudin (*left*) and William Dampier (*right*); the latter was removed by Robert Brown from a specimen at Oxford University Herbarium. (Natural History Museum, London)

specimen and realized he was mistaken: the plant had been previously named. Unfortunately, he had already told the honoree, and Darlington was so thrilled he showed specimens to his friends at a state fair. When Torrey wrote to Darlington of the problem, he took a mild swipe at Gray: "It is surprising that Dr. Gray . . . did not find this out."[20] Also, Torrey complained to Gray of the poor quality of some of the specimens collectors sent him, a persistent issue with botanists who had little experience collecting in the rugged wilderness, which offered endless challenges.

Fortunately for Torrey, the same package contained a solution to the problem of naming a plant for Darlington: a pitcher plant specimen from near Mount Shasta in northern California. On the Wilkes Expedition, the plant's stems and pitchers, which are modified leaves, were collected, but there were no flowers, needed for a proper description. Now Torrey was thrilled—he had flowers as well as leaves. The pitcher was hooded with two projections from the hood, making it clearly different from other North American pitcher plants (figs. 9.3a–c). Torrey quickly wrote to Gray, sending him a specimen so that Gray's artist, Isaac Sprague, could produce an illustration. Torrey later carefully annotated the drawing, pointing out features that needed slight alterations. The plant was considered so noteworthy that its description was printed in its own booklet by the Smithsonian Institution.[21]

Darlington was grateful, writing to Torrey that it was a wonderful honor, but he was gun shy: would he end up losing this plant as well? He knew that the "inexorable" George Bentham had recently described a new pitcher plant species from New Guinea; could this be the same as his new *Darlingtonia*? He had not seen Bentham's publication so was in doubt. With that publication in hand, Torrey wrote right back, assuring Darlington that his genus was solid. To make the differences clear, he went to the trouble of tracing the illustration of Bentham's plant, *Heliamphora nutans*, copying out the legend as well.[22]

Managing Collectors

American plant collectors were often in the military, attached to such operations as the Mexican-American War and later the United States and Mexican Boundary Survey. Botanists saw these missions—even the war—as opportunities for collecting in new areas. Solitary travel was difficult and could be dangerous, so military backing was important. Later, railroad surveys served the same purpose, providing an infrastructure for collectors. The Civil War obviously slowed the pace of collecting, but it picked up again in the 1870s, with the advent of railroads in the West making more terrain accessible.

George Engelmann was a German physician who immigrated to St. Louis and eventually collaborated with wealthy businessman Henry Shaw, who established the Missouri Botanical Garden. Engelmann became an expert on cacti and made several collecting trips west, but he also relied heavily on collectors. Asa Gray often teamed up with him to manage them, sharing expenses and specimens. Good plant hunters were not easy to find and required careful attention: providing guidance as to what and where to collect, making sure they had needed supplies, and keeping them on track while not pushing them too hard. Collectors had to be physically fit to deal with difficult terrain, have enough botanical knowledge to know what they were searching for, and make good-looking specimens with flowers and/or fruits. Some collectors, like Charles Parry, a physician who served with the United States and Mexican Boundary Survey, and William Emory, an army engineer, were particularly useful because they stuck to the job for years.[23]

Gray complained about the number of specimens of common weeds like lamb's quarters that he received, noting that many collectors did not seem to realize it was a non-native plant, introduced by early colonists. A common complaint from those in the field was that their clients didn't appreciate the difficulties of bad weather, unreliable transportation and supplies, difficult terrain, and locals who did not necessarily value botany. They had to make ends meet by selling limited-edition collections with printed labels, called exsiccatae (exsiccati for fungi), though it was a great deal of work creating them and definitely kept collectors occupied in the off season.

Not everyone traveled long distances to find interesting plants. Some collectors, like the physician Gideon Lincecum, explored near home, learning about medicinal plants from indigenous peoples. He, like other plant collectors, found that native healers were eager to share information.[24]

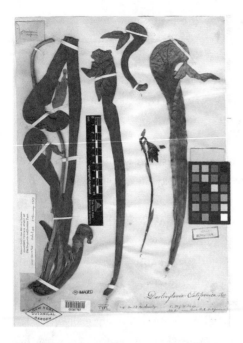

Figure 9.3a. Type specimen of the cobra plant (*Darlingtonia californica*). (Image courtesy of the C. V. Starr Virtual Herbarium [http://sweetgum.nybg.org/science/vh/])

Figure 9.3b. *Darlingtonia californica*: photograph of live plant by David Berry.

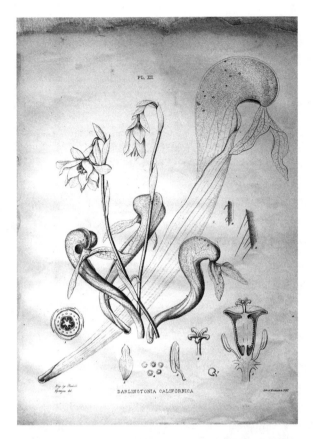

Figure 9.3c. *Darlingtonia californica*: illustration by Isaac Sprague in Torrey's *On the Darlingtonia Califor-nica*. It displays Sprague's ability to depict a plant accurately, guided solely by dried specimens. (Missouri Botanical Garden, Peter H. Raven Library)

William Henry Ravenel, a South Carolina plantation owner and amateur botanist who was bankrupted by the war, attempted to eke out a living from botany. Before the war, he had published a series of fungal exsiccati. After the war, he marketed what he had left, as well as selling his microscope and books.

Ravenel also resumed correspondence with his fellow botanists in the North, including Asa Gray, who offered advice and assistance. A Philadelphia nurseryman gave Ravenel a "loan" that he did not have to repay, and a botanist in Massachusetts helped him sell his fungal specimens. Friends in Washington, DC, arranged for him to meet the commissioner of agriculture, who contracted with Ravenel to collect specimens and seeds, including on a trip to Texas. Just as Joseph Banks saw no reason to sever contact with French botanists, those of the North and South let plants heal the wounds of war.[25]

As travel to the West improved, the scope of collecting broadened. The American Museum of Natural History (AMNH) in New York created a forestry exhibit early in its history, some of which is still on display today. Many of the wood specimens were procured by the Vermont plant hunter Cyrus Pringle during plant-collecting trips to the West Coast and the Southwest. His work was directed by Charles Sprague Sargent of the Arnold Arboretum in Boston who was collaborating with the AMNH and working on a survey of American trees. Sargent's aim in helping to create tree exhibits was to focus attention on the nation's forests and the need to preserve them. The writer and diplomat George Perkins Marsh had made Americans aware of the riches they still possessed in their forests and of how fast these were disappearing. He had spent time in Europe and knew how little forest land remained there; he warned of a similar fate at home.[26]

Pringle's task was to find, haul, and pack for shipment representative timber from Southwest species: five-foot-long cuts through the trunk as well as cross sections. He arrived in Tucson, his base, on one of the first trains into the area, as railroad access opened up remote locations to collecting and made shipment of massive specimens possible. On his way west, Pringle visited George Engelmann in St. Louis, who wrote to Sargent: "He [Pringle] has queer instructions from you! To get trunks of well grown trees 5 feet long, even if 2 or 3 feet diam. on muleback, if no wagon reaches to the localities. I asked him whether he might not get sections; he thought not. Well you have to be more definite in your instructions and he must act according to circumstances."[27]

The pace of collection was also a problem. Pringle went to California in July with Sargent expecting him to find, collect, and ship specimens of forty tree species, all before the fall blooming period in Arizona. Once cut, huge trunks then had to be carefully boxed since the pieces were for exhibit.

Sargent complained repeatedly of Pringle's failure to meet deadlines and even withheld payments. Pringle eventually refused to work for Sargent, devoting himself to collecting plant specimens in multiples and selling these as exsiccatae. Despite these collecting difficulties, forestry and other botanical exhibits became common in the late nineteenth century as one manifestation of the public's interest in natural history.

Natural History and Botany

REASONS FOR THE INTENSE INTEREST IN nature that characterized much of the nineteenth century in the United States and Europe shifted over time.[1] Early on, the continuing popularity of Linnaean classification made natural history accessible to a larger audience. There was also a moral undercurrent through natural theology: the belief, particularly in Britain and the United States, that studying nature was a way to learn about God. Exploring some part of the natural world was, in truth, a good thing to do. The mental activity involved in learning about organisms was good for the intellect. Being active outdoors, investigating nature in situ, was healthy as well, and one way to document all this good living was by collecting specimens.

Some collectors were casual, simply slipping cuttings into field guides at the pages describing the species, sometimes with notes on where the plant was found or with a sketch. Others were particularly reflective in their use of herbaria. As already noted, the practice appealed to philosophers and literary figures such as Jean-Jacques Rousseau and Johann Wolfgang von Goethe. The philosopher John Stuart Mill was interested in botany throughout his life and saw the hierarchical classification of living things as a model for ordering many aspects of human affairs, such as the law. He collected so extensively that his daughter-in-law outfitted a room for his specimens.[2] One expression of the writer John Ruskin's growing interest in plants was the herbarium he created during one of his trips to the Swiss Alps.

As the century progressed, there were major economic forces at work that allowed greater access to new experiences of nature even as urbanization separated many from constant interactions with it. The industrial revolution and especially the steam engine's invention meant the advent of steamboats and railroads that made travel easier and ultimately cheaper, providing greater access to collecting areas. Steam also powered printing presses, making for cheaper books, and new paper-manufacturing techniques using wood pulp did the same. Glass and cast-iron manufacture enabled construction of large glasshouses. Joseph Paxton, who designed the Crystal Palace for London's Great Exhibition of 1851, was a gardener by profession. He had begun by building glass-walled conservatories for his wealthy patron, the Duke of Devonshire. In one of these he managed to coax into bloom the massive *Victoria* water lily, one of the species discovered in the nineteenth century that fueled a passion for plants.[3]

Rising literacy rates and the increasing availability of newspapers and magazines also spurred interest in natural history. More educational opportunities were available, and female as well as male authors produced science books aimed at women and children. Among the most noted in the United States was Almira Hart Lincoln Phelps, whose *Familiar Lectures on Botany* went through seventeen editions. One of its readers was the poet Emily Dickinson, who took a serious interest in plants and created her own herbarium. While at school, Dickinson wrote to a friend: "Have you made an herbarium yet? I hope you will if you have not, it would be such a treasure to you; 'most all the girls are making one."[4] Pasted onto sixty-six pages, there are 424 specimens in her bound collection, all but 60 with Latin binomials (fig. 10.1). This is not a work of science: there are no place names or dates of collection, nor was there an effort to arrange related species together. However, it is a carefully composed book and hardly unique. Thousands of amateur herbaria from the nineteenth century still exist in private collections, museums, and libraries.

Pressing plants was a relatively easy technique to learn and provided an attractive and permanent record of an encounter with nature, so it's not surprising that herbaria were particularly popular at this time. There were several reasons for making such collections, including the social aspect Dickinson emphasized: one did not want to be left out of the collecting circle. Herbaria were sometimes part of a school curriculum; there were

Figure 10.1. Page from Emily Dickinson's herbarium, circa 1839–46. (MS AM 1118.11, seq. 19. Houghton Library, Harvard University)

printed notebooks with places to attach pressed plants and lines to fill in the collector, date, location, and species. An herbarium could be used as a learning tool and a memory tool: to remind an amateur botanist of what a species looked like so he or she could compare it to a new find.

An herbarium was more than simply a kind of botanical stamp collecting, of documenting the plants a collector had "captured." It was a way to learn about plants and internalize that knowledge, with tactile and visual experiences reinforcing the cognitive. Collectors were not just looking at plants, though obviously they did that closely. Perhaps noting a lingering scent, they were also clipping, arranging, and pressing material, then referring to it later to solidify their understanding. Researchers have noted that quiet time spent doing what are apparently menial laboratory tasks provides important contributions to scientific inquiry. In an examination of still-life painting, the poet Mark Doty writes that when we describe the world in words or in a painting we come closer to saying what we are; the process allows us to think things through; the object of description is not separate from us.[5] Making herbarium specimens is a similar blend of art and science, and the pleasure of the work is part of the attraction of natural history.

Collectors could be omnivorous, amassing any kind of specimen they could find. Buying up others' herbaria, the wealthy could afford to create huge collections, as did two men in positions that put them in contact with many collectors. Manchester shipping executive Charles Bailey and his friendly rival James Cosmo Melvill, secretary of the East India Company, decided early on that their activities should not conflict since they both intended to donate their specimens to the Manchester Museum, where these collections still reside. Bailey focused on the British Isles, Europe, and Africa, while Melvill took on the rest of the world. Bailey's collections eventually amounted to over three hundred thousand items, and Melvill's, not surprisingly, were of similar size. There were also collectors at the other end of the economic spectrum, particularly in Britain. Many in the working classes used their leisure time to collect, often getting together afterward at a pub or tearoom to share specimens and information. They frequently sent specimens to wealthy collectors in exchange for information and sometimes for payment. Botany, at least to a point, spanned class distinctions.[6]

Areas of Interest

Some collectors had focused goals: attempting to locate all the species in a geographic area or all the representatives of a particular family or genus. Collecting and breeding orchids was a mania among the wealthy, especially after greenhouses became more common. Plant hunters specializing in this family often cultivated the plants near where they were collected and sent back not only specimens but live plants since orchids were often difficult to propagate outside their native ranges. As well as pressing specimens, plantsmen would also draw or have drawings made of the plants to advertise what they had on offer. Orchid fans took great pride in their ability to nurture these plants. The businessman John Day became so devoted to the family that he created fifty-three scrapbooks of notes and watercolors of his plants and others that he drew at Royal Botanic Gardens, Kew. If he could not identify a species, he sent a drawing to the noted orchid botanist Heinrich Gustav Reichenbach in Germany, who described many new species in part based on Day's watercolors.[7]

At midcentury there was a fad for ferns, fed by variations on the Wardian case making it possible to grow some species at home in enclosed glass cases. Fern herbaria likewise became common. For those who wanted the specimens without bothering to traipse through the countryside, scrapbooks filled with labeled specimens were sold by companies specializing in natural history materials, another phenomenon of the time. The specimens could be common ferns or exotic ones. New Zealand was particularly noted for these albums because ferns were a common feature of the land. Collectors catered to well-heeled clients around the world (fig 10.2). To have such an exotic volume was the mark not only of wealth but of botanical sophistication.[8] For the more hands-on, natural history suppliers sold equipment, from plant presses for drying specimens to hand lenses for examining them. Marks of a serious collector were a lens and a vasculum, a tin box with a leather strap attached to sling over one's shoulder, to hold cuttings.

Other factors in natural history's popularity were the growing availability of leisure time and improved transportation: railroads and bicycles made it easier for city dwellers to access nature. The concept of the vacation developed as the century progressed. The seaside was an attractive venue, opening up the world of algae to the nature-hungry. Seaweed became a rage, with shops in coastal towns like Plymouth in England and Newport in Rhode Island offering scrapbooks in which to press algae. The technique used was different from that for land plants. The specimen was floated in a

Figure 10.2. Pressed fern album cover with wood inlay work, circa 1875, Auckland, by Thomas Cranwell and furniture maker Anton Seuffert. (Purchased 1988 with Charles Disney Art Trust funds. CC BY-NC-ND 4.0. Museum of New Zealand Te Papa Tongarewa [GH003578])

pan of water, then a sheet of paper was placed underneath it and gently raised to catch it. The results were often beautiful and, particularly in England, professionals produced albums of specimens for sale in coastal areas. Herbaria were definitely an aspect of travel in the nineteenth century. They were popular because they allowed a closeness to nature even after the fact. Printed albums with specimens from the beaches, the mountains, or even the Holy Land were not only remembrances for those who traveled but also gifts for those who could not, uniting the scientific with the aesthetic and the religious in an attractive way: what more could one ask of a souvenir?[9]

Many of the amateur herbaria, especially for seaweeds, were hardly scientific. Algae specimens were frequently left unnamed and were often presented in decorative arrangements (fig. 10.3). More scientific were the cyanotypes created by Anna Atkins, a photography pioneer whose images of algae were all labeled, and writer Margaret Gatty's specimens were named according to the definitive guide to British algae.[10] Presentations of plant and alga specimens can be placed along a spectrum from art to science and also from the plant itself to its representation.

Another collector's item was the nature print, in its simplest form made by inking a flattened plant specimen and pressing it between sheets of paper to create an impression, attempting the closest possible representation of an actual specimen. Though the practice extended back to the early modern era, techniques were developed over time to improve the prints and make them in multiples. In the eighteenth century, two German botanists published books describing plants illustrated with nature prints enhanced with watercolor. However, since the plant itself was used in printing, only a few copies could be made before the specimen disintegrated. In the nineteenth century, Alois Auer, director of the Austrian National Printing Office, developed a method for passing specimens through a rolling press between plates of polished lead and copper, leaving a clean impression in the softer lead from which a printing plate could be created. Auer produced a multivolume flora of Austria employing this method. A similar one, which Auer claimed was stolen from him, was used by the British printer Henry Bradbury to create beautiful volumes of fern and of algae prints, specimens particularly well adapted to nature printing since they don't produce flowers.[11]

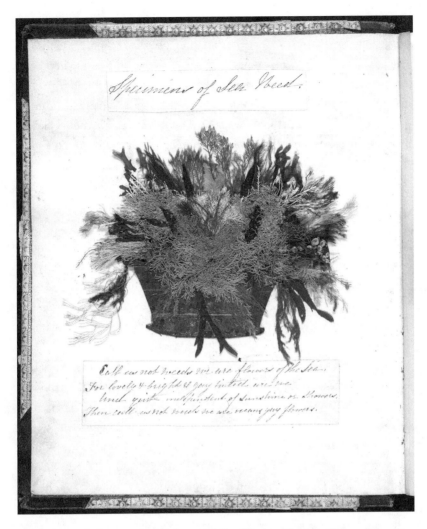

Figure 10.3. Title page of *Specimens of Sea Weeds,* a British album, circa 1840. (Yale Center for British Art)

This was a high point for nature printing. Costs were prohibitive and the results varied with the type of plant. The prints were more artistic than scientific. It was difficult to print the details of flowers that are so important in plant taxonomy and to provide the level of information found in herbarium specimens or good illustrations. New printing methods like lithography reduced costs and eventually allowed for color reproduction. Cheaper paper made small-format guidebooks popular, often illustrated with wood engravings. Later, photographs appeared in field guides.

Institutionalization of Botany—and of Herbaria

At a time when natural history was becoming ever more popular, professional botanists were moving away from it, developing the discipline of botany as separate from the more general field. Paid positions in botany became more available; Joseph Dalton Hooker at Kew and Asa Gray at Harvard are examples of those who depended on employment in the field for their livelihood. This had not been possible earlier, when most of those interested in plants were either independently wealthy or practiced medicine or pharmacy. However, with professional positions came not only responsibility to do taxonomic work but administrative tasks as well. Hooker and Gray often apologized for delays in answering letters because of their workload, and the problem was real.

Joseph Hooker had taken over at Kew in 1865 after his father William Jackson Hooker died. The elder Hooker had built up a large herbarium during his career, moving it from the University of Glasgow, where he had taught, to Kew when he assumed the directorship there. At the time, a plant collection was often seen as the property of the collector, acquired by gathering the plants, buying specimens, or trading duplicates for species not already in the collection. Amassing specimens in multiples, when possible, became more common as a source of duplicates for trade or sale, though rarity or transport problems might make this impossible. William Hooker's collection went to his son, who sold it to Kew; it became part of a collection that today is the world's largest, with over 8 million specimens.[12]

Kew eventually extended over 299 acres, including Kew Palace, forested land, conservatories, a pond with water lilies, art galleries, a rock garden, and even a pagoda. However, right before William Hooker took over, it was close to being destroyed. Joseph Banks had nurtured Kew and its role in international plant collecting, but after his death in 1820 the garden deteriorated, and land so close to a growing London was considered desirable for other uses. Leaders in the botanical community finally convinced the government of Kew's historical and botanical value, as well as its economic potential for growing plants from the colonies for possible introduction into agriculture or horticulture.[13]

Once William Hooker arrived at Kew in 1841, the garden's future was assured as a place of science and beauty. As far as Hooker was concerned, science was at the fore, and he saw his mandate broadly as including documenting useful plants just as Banks had done in starting to preserve plant products, what he called an economic botany collection. Growth of this collection was in part the result of Kew acquiring the plant collection of the East India Company's museum in London, including thirty-six tons of wood specimens, an herbarium, and plant products from grass mats to wood carvings. This was a result of the company's dissolution in the aftermath of the Indian Rebellion of 1857. When examined, the company's collections, made by such notables as Nathaniel Wallich, were found to be poorly preserved and in some cases rotting, an inevitable consequence of an herbarium not receiving attention over a period of years.[14]

A later threat to Kew, specifically to its herbarium, arose in the early 1870s as the British Museum's natural history collections, including plants, were being moved to a new building that would become the Natural History Museum, London (NHM). The director, Richard Owen, argued that it made no sense for the city to have two herbaria; the new museum space could accommodate the Kew collection, with the herbaria's specimens integrated and used more effectively at NHM. Eventually, Joseph Hooker was able to stave off the attack, citing several reasons, including the mundane but significant one that Kew and the museum used different sized sheets, so merging them would have been difficult, requiring much cabinet refitting. To address some of the overlap between the collections, in the 1960s Kew and NHM each agreed to focus future collecting in specific geographic areas and divided responsibility for different categories of nonflowering plants between them.[15]

American Institutions

As the Hookers built Kew, Asa Gray created a place for botany at Harvard as its first professor of natural history, with the administration agreeing that he would teach only botany. Gray wrote a textbook to accompany his lectures and later published several others. He also took over the botanical garden located next to the house provided for him. He brought his herbarium with him and enhanced the library, a must for taxonomic work. They both became the foundation of Harvard's botanical collections. By the time Gray retired in 1873, he had accomplished a great deal with little assistance. Since he was the only botanist, all burdens of administration fell to him. It was only in 1870 that he was able to begin hiring the next generation. Institutionalization required human capital to maintain it. His four successors each assumed a portion of his work, which had included managing the herbarium, library, botanic garden, and economic botany collections; the last became a botanical museum.[16]

New York Botanical Garden (NYBG) was established in 1891 with botanist Nathaniel Lord Britton as the first director. His wife Elizabeth, an expert on fungi, was also involved in the garden's development and research. Britton managed two major taxonomic projects that required much collecting: a series of volumes on the cactus family and another on the flora of Puerto Rico and the Virgin Islands. In 1915 alone, forty-three thousand specimens were acquisitioned by the garden. The Brittons also bought European collections, acquired Columbia College's herbarium, which included John Torrey's specimens, and added their own specimens to form the foundation of NYBG's herbarium.[17]

California was also maturing botanically, though not without challenges. On the morning of April 18, 1906, a magnitude 8.3 earthquake damaged not only buildings but water mains and gas lines in San Francisco, creating conditions for a massive fire. The California Academy of Sciences (CAS) herbarium curator Alice Eastwood, in anticipation of such an event, had sorted out the type specimens, those that had been used in describing new species. She deposited them in a portable cabinet that could be lowered through a window if necessary. When she and a friend reached the building, they found the marble staircase leading up to the herbarium was shattered, though its iron railing was intact. They scaled the railing and reached the sixth floor, but the type cabinet had been damaged and couldn't be moved. They put specimens into a work apron tied with cords; Eastwood went down to the street while her friend lowered packages of specimens in the apron until all the types were retrieved. They took the almost fifteen hundred specimens to Eastwood's apartment, not an easy task amid the chaos. As the fire spread, the cache had to be moved twice more; eventually Eastwood stored it in a bank vault.

Eastwood's home and most of her possessions were destroyed in the fire, but it was the type specimens she most valued and managed to protect.[18] Her forethought in sorting them out made their rescue possible and explains why types are kept separately in many institutions. The CAS building itself did not survive the fire, nor did most of the specimens collected over fifty years in what then composed the largest herbarium in the western United States.

The CAS was founded soon after the California gold rush, its emphasis on natural history a symbol of stability and intellectual development in what was then a boomtown. Only thirty years earlier, on the Wilkes Expedition, *Darlingtonia californica* had been collected in the mountains of northern California before the area had even become part of the United States. By the time of the fire, there was not only an herbarium in the CAS but one in an active botany department at the University of California, Berkeley, and another at nearby Stanford University. This substantial botanical infrastructure was mirrored in other parts of the United States. As the twentieth century began, the momentum for collecting and studying plants continued, with herbaria expanding at universities, botanical gardens, and museums throughout the country.

Also influencing the development of botany in the United States was the Morrill Land-Grant College Act of 1862, which provided land for states to create colleges focused on agriculture, science, and engineering. These became important loci for plant collections and remain so to this day. Earlier, in 1815, the British Parliament passed a bill requiring all medical students to take a course in materia medica, essentially medical botany. These

two educational initiatives helped to institutionalize botany and to separate
it from natural history, which during its heyday grew so popular with the
general public that it seemed to drift further and further away from "real"
botany. By 1900, natural history became part of the school curriculum, and
many argue that this led to its waning popularity: it was no longer considered fun.[19] Also, the Darwinian theory of evolution weakened the link between biology and religion, so nature study as being good for the soul
became a less popular idea.

Surveys and Museums

Another aspect of botanical institutionalization in the late nineteenth century and a major spur to herbaria was the development of surveys, many
government sponsored, to systematically collect in particular areas. The
great expeditions earlier in the century were wide ranging, but they resulted
in rather haphazard collections in relatively unknown areas. Gathering was
done when time was available and habitat worth investigating; plants might
or might not be in flower or fruit, yet these were important for species identification. Surveys, on the other hand, were both intensive and extensive;
collectors often visited the same locales repeatedly to ensure that all specimens were represented by flowers and/or fruits.[20] The United States Biological Survey organized nationwide, though most botanical surveys were
done on the state or local level. These enterprises were less about discovering new species and more about inventorying what was growing in a particular place at a particular time, the beginning of a more scientific approach
to collecting that eventually led to biodiversity research.

Surveys were the result of several trends in the United States' economic, scientific, and cultural development. The creation of an extensive
railroad system, particularly after the Civil War, made large portions of the
country accessible for teams of collectors to travel economically. Roads
were also extended and improved. Urban and industrial expansion brought
habitat destruction that alarmed many who sought to document these
changes as well as the nature remaining in undeveloped areas. At the same
time, land grant colleges enlarged their offerings, with botany as an impor-

tant component of agricultural programs. The Nebraska survey was spear-headed by Charles Bessey, who taught botany at the University of Nebraska. It began as a student project, a way to highlight the importance of the educational system to the advancement of research.[21]

The scientific nature of surveys was emphasized by the use of printed forms and field notebooks to record information uniformly. The specific date of collection was necessary—"Sept 1867" was not good enough—and the collector had to be named. Usually collectors recorded a number for each plant, something that Alexander von Humboldt had done but was an uncommon practice until much later in the century. Locality information also became more exact, going beyond "near Tucson" or "on the Colorado River." These practices led to more informative specimen labels, a boon for those using these specimens in today's research. Despite the valuable data they produced, by the 1930s surveys became less common, in part because natural history became less appealing to the public than it had been in the nineteenth century. Lack of interest meant less funding. From that time on, collecting was more focused on smaller areas or on particular plant groups. However, the legacy of surveys remains in the rich collections they produced, which continue to inform botanical inquiry.

One aspect of the natural history boom that continued into the twentieth century was the development of natural history museums. Many featured herbaria as well as zoological, mineral, and anthropological collections. The Field Museum in Chicago and the Smithsonian's National Museum of Natural History amassed particularly large herbaria, while New York's American Museum of Natural History had none, having entered into an agreement with NYBG to make the latter the major plant repository for the city. Museums were flourishing, riding on the building boom in impressive structures that had begun in the second half of the nineteenth century, and their twin goals of research and education often reinforced each other. This was especially true for the development of dioramas, signature exhibits of the first half of the twentieth century. In their most sophisticated form, dioramas have painted backgrounds of specific habitats with lifelike taxidermied animals and equally realistic plants made of paper, wire, wax, and other materials.[22] These landscape replicas allowed city dwellers with limited resources to experience something of the natural world, and they continue to serve this purpose. Many museums retain some or all their dioramas because of their popularity. Even in the video age, it is appealing to see animals, albeit not living ones, up close within their natural habitats.

Though most visitors tend to focus on animals rather than plants in dioramas, exhibition designers worked hard on their authenticity, making full use of available research collections. Much work went into planning and creating dioramas, with herbaria integral to the process. On expeditions to areas of interest, botanists and zoologists collected specimens for their research and as reference material for the displays. An artist was usually part of the team, but the botanist B. E. Dahlgren, a Field Museum herbarium curator, played both roles, not only collecting specimens but also painting watercolors of plants that would be replicated in a museum laboratory created for the purpose. These drawings were often filed in the herbarium along with specimens after they had been used in creating lifelike plants. At times the lab temporarily relocated to south Florida so that artists would have live tropical plants readily available as models.

The plants were used not only in dioramas but in the Field's extensive economic botany exhibits, though it was in botanic gardens that economic exhibits were most extensive. Taking the lead from Kew, many botanic gardens had museums for their economic botany collections of tools, fabrics, medicines, foods, and other items made from plant materials. Kew extended exhibits until they were spread over four buildings, with two for displays on trees and their products. These were very much about the riches of the British Empire, and other colonial powers followed suit, with Belgium founding a forestry museum around 1900 to present materials from its African colonial holdings.[23] But not all forestry exhibits were about economics. One particularly memorable item at the NHM is a cross section through a giant sequoia, with dates marked on the rings to indicate significant events in human history and to emphasize its age; it remains a powerful visual representation of longevity and size. Similar exhibits are found in other natural history museums.

Forest exhibits were often backed up with herbarium xylaria, wood collections. They provided reference material when lumber or wooden artifacts were brought in for identification and are still used today in determining the species in archaeological finds and when timber from rare species has been illegally harvested and imported. Important to this work is another element in xylaria: collections of thousands of wood tissue microscope slides for studying wood anatomy. These were made by botanists studying the structure of wood and how it developed. When interest in the field waned in the 1970s and 1980s, these and other wood-related collections seemed to be just taking up space. They were among the victims of herbarium decline, often discarded or sent to institutions willing to accept them. But interest in specific scientific areas varies over time, a good reason for preserving collections. At Oxford University, a set of slides with thin sections of fossilized plants was literally pulled from a dumpster and later discovered to contain the only known evidence of a particular type of root structure found in a 310-million-year-old specimen.[24]

The survival of this material hints at the survival of natural history as a discipline, as will become apparent in future chapters. The idea that all nature is related and must be studied as such has garnered increased interest in the twenty-first century as the stresses on the natural world continue to increase. But this recent resurgence was preceded by a period when other areas of botany overshadowed taxonomy before its central importance was again recognized.

ELEVEN

Evolution and Botany

UNLIKE JOSEPH HOOKER AND ASA Gray, Charles Darwin was an independent naturalist, not affiliated with any institution. He could manage this because he had inherited a substantial income and added to it by astute investment. He was in the tradition of gentlemen naturalists, who were no less rigorous in their work or expertise because they lacked institutional ties. Darwin was a member of learned organizations, most notably the Royal Society and the Linnean Society. He developed an extensive network that included the likes of geologist Charles Lyell and philosopher William Whewell, who coined the term *scientist* for those interested in any of science's disciplines. Its use relegated the word *naturalist* to amateurs. Charles Darwin was both a naturalist and a scientist. He caught the fervor for natural history early in life, collecting insects and gardening with his father, a physician. His grandfather Erasmus Darwin, also a physician, wrote poetic works on botany, including one on Linnaean classification, and suggested the idea of species change.[1] So plants were an important part of Charles's life from the beginning.

Species Variation

Darwin's work is so multifaceted and rich that there are endless ways to examine it, including from the viewpoint of herbarium specimens. While enrolled at Cambridge University, he attended John Stevens Henslow's botany lectures. Especially during the latter part of his education, the pair explored the countryside collecting plants and insects. Henslow kept an herbarium, and Darwin contributed to it. The way Henslow arranged his specimens was a little different. A recent analysis found that two-thirds of his sheets in the Cambridge University Herbarium are what he termed "collated," that is, having more than one specimen of a species. In many cases, specimens on a sheet included ones from different collectors, dates, and locations. This may not seem noteworthy, especially for small plants, but their placement is interesting. They are often arrayed in height order (fig. 11.1) or with the largest specimen in the middle, with plants of descending height on either side—arrangements emphasizing variation.

Around the time he taught Darwin, Henslow was investigating variation within plant species.[2] He grew primulas, varying the amount of moisture, manure, and shade, and noting that they differed in some ways from those growing in fields. Earlier, he had made drawings of the different lengths of reproductive structures that influenced pollination in the cowslip, *Primula veris,* something Darwin later investigated. The argument proposed in an analysis of Henslow's specimens is that they, coupled with observations on variation presented in Henslow's lectures, became part of Darwin's "mental architecture." They were so integrated into his thinking that he might not have even realized the debt he owed Henslow, beyond the fact that this mentor connected his former student with Captain Robert Fitzroy of the *HMS Beagle,* leading to Darwin's five-year voyage around the world.[3]

Darwin had an opportunity to collect many plants on the *Beagle* expedition, hundreds of them, which he sent to Henslow along with animal skins, fossils, and other items. After sending the first shipment he had to wait two years for a letter from Henslow to catch up with him. In the meantime, he worried that no correspondence meant Henslow was displeased with the material. It was a relief when Darwin finally learned that Henslow was grateful for the specimens, although he commented that Darwin should not send scraps. The entire plant needed to be included whenever possible—leaves, roots, stem, flowers—and one of the leaves should be turned back to reveal the underside. Also, it wasn't necessary to sew down the specimens; they traveled better when left loose.[4]

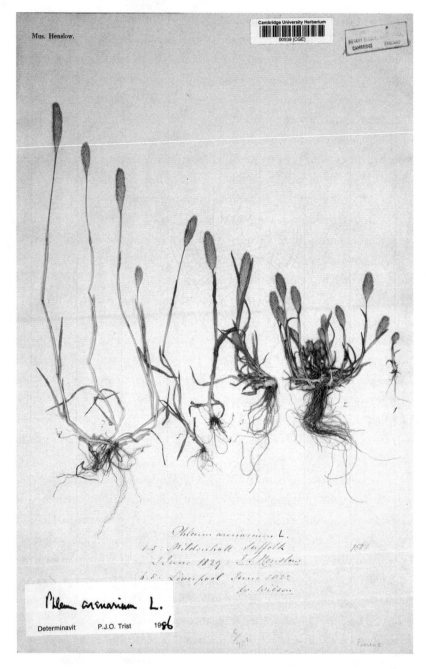

Figure 11.1. Specimens of sand timothy (*Phleum arenarium*) arranged in size order, collected by W. Wilson and John Henslow. (Cambridge University Herbarium [CGE])

Soon after he returned to England, Darwin delivered the last batch of specimens to Henslow, who agreed to begin identifying them. This turned out to be a slow process, in part because the assignment coincided with Henslow's move fifty miles from Cambridge. He remained a Cambridge professor but usually visited there only to give his lectures. Another problem was that many of the plants were unfamiliar to Henslow, who did not have much experience with tropical species. Darwin kept prodding him for several years until Henslow turned the specimens over to the young botanist Joseph Hooker, who had just returned from a round-the-world expedition with the British Navy. Hooker found that many of Darwin's plants from the Galápagos Islands were endemic to the islands, found nowhere else. In some cases, a species occurred on only one island, information that Darwin was reassured to learn since it fit with what he had been told about his bird specimens.[5]

At this time Darwin was working on "the species problem," that is, how species came to be. If a deity was not responsible for the multiplicity of living things, what was? The variation he saw within species provided a clue, as did an unlikely source, the economist Thomas Malthus's essay arguing that population rise among lower classes would inevitably lead to food shortages, resulting in many deaths among this social stratum. Several days after reading this, it occurred to Darwin that something similar could occur in nature. Within a species, those with particular traits might be better able to survive difficulties and produce offspring, and their traits would become more prevalent in future generations. Particularly if the group was small and isolated, as on an island, this could lead to a population so different from others it would be a distinct species. This idea occurred to Darwin in 1838 but it took him years to amass information to support it. When Alfred Russel Wallace wrote him from the Malay Archipelago in 1858 describing a similar explanation, Darwin was forced to make his theory public in order to share recognition with Wallace. A scientist is credited with work only when it becomes public knowledge.[6]

During the intervening years, two botanists were among Darwin's closest confidants. Darwin asked Joseph Hooker to examine his plants from the Galápagos because Hooker had also visited Pacific islands on the

expedition from which he returned in 1843. They became so close that
when Darwin wrote a 230-page summary of his theory in 1844 and gave it to
his wife, he requested that in case of his death, she ask Hooker to arrange its
publication. At one point Hooker bluntly suggested that while Darwin's
interests were definitely broad, extending from variation in domesticated
animals to fossils and plant breeding, he had not delved deeply into any one
group of organisms. He needed to study some segment of the living world
closely enough to understand what distinguishes one species from another.
This triggered Darwin's eight-year odyssey studying barnacles, which resulted in a two-volume work.[7] After sorting through all these species, trying
to differentiate among them and determine the traits distinguishing them,
Darwin had a firmer grasp of what his term for evolution, *descent with mod-
ification,* meant—what modifications were significant enough to distinguish among species.

Through the 1850s, Darwin didn't focus exclusively on barnacles; he
had a broad array of correspondents providing him with information on
many topics. That is how he befriended Asa Gray of Harvard University.
They had met briefly during Gray's first trip to London, and Darwin mentioned this connection in a letter to Gray when he asked about the range of
alpine or mountain plants in North America, since Gray had listed them in
one of his books.[8] Like many naturalists of the time, Darwin and Gray were
influenced by Alexander von Humboldt's work on biogeography. They
were also empiricists who considered direct observation and experimentation key to scientific inquiry. Their correspondence continued, often becoming a three-way conversation including Joseph Hooker, whom Gray
had also met in England.

In 1853, while U.S. Navy commodore Matthew C. Perry was in Japan
negotiating a treaty opening the country to trade, two Americans associated
with the mission, one a friend of Gray's, made a small plant collection and
sent it to him. What struck Gray about these plants, as well as other Japanese species he had encountered, was how similar they were to those of the
northeastern United States. When he had examined a wide enough selection, he arranged them in a table and found that of 580 Japanese species,
many more were akin to eastern North American plants than to those of

western North America or Europe. Considering this odd result, he posited that the plants in these two areas had a common ancestry, but due to fluctuating climates, which remained most similar in eastern North America and Japan, the species were better able to survive in these regions.[9] Like Hooker, Gray also provided Darwin with botanical information for the many plant studies he undertook after publication of *On the Origin of Species.* They experimented in their gardens on, for example, the twining of vines, and Gray sent seeds and seedlings of American plants.

The Type Specimen

For most nonscientists, the concept of species is usually clear-cut: a lily is different from a rose and both differ from a magnolia, but for biologists who study finer distinctions among organisms, the species idea can be problematic. Linnaean nomenclature brought some stability by providing a standard naming structure, the binomial of genus and species, and his list of species became the foundation for modern taxonomy.[10] Yet as more and more species were discovered, the list became unwieldy, especially as the same plant was sometimes named by more than one botanist, unaware of another's work.

Linnaeus, like most botanists of his day, saw species as fixed, the result of divine creation, and assumed that the genus, a group of species with similar traits, was also a natural category. However, broader classifications were a different story. Linnaeus admitted that his classes were based on a plant's reproductive organs because these were easy traits to observe and quantify. But the artificiality of Linnaean classification bothered many botanists, so in the late eighteenth into the nineteenth century several devised natural classification systems. Those of Antoine de Jussieu and Augustin de Candolle were particularly well-developed, based on collections of traits rather than just on reproductive structures.[11] In some cases it was obvious how plants in a number of genera were similar enough to be placed in the same family; in others, the divisions were less clear-cut and seen differently by individual taxonomists, depending on the traits they deemed important. Still, the comparative work continued, and families were grouped into orders and then classes to form a multi-tiered hierarchical system.

With Charles Darwin's work on evolution, natural classification was given a firmer theoretical foundation. Darwin argued that all organisms evolved from a single primitive life form, so all organisms were related. There was now a biological explanation for why some plants were more similar than others. Those that were alike were assumed to be more closely akin through descent from a common ancestor. Taxonomy—the identification, description, and classification of species—continued in much the same way as in the past, but now there was more emphasis on systematics: working out relationships among species in terms of evolution. It's not surprising that herbarium specimens were pivotal in this work and a reason why large collections with a wide variety of species and genera were essential. Antoine de Jussieu worked at the large Paris herbarium and Augustin de Candolle was first at Montpellier, long known for its botanical research, and then in his native Geneva, where he built a significant collection. This was used by his son Alphonse, who continued work on his father's multivolume description of his natural classification system. Alphonse became one of the most prominent botanists of his time. At the first international congress on botany in 1867 he proposed a taxonomic code, a set of rules for naming plants and stabilizing classification.[12]

The code was a facet of the institutionalization of botany. Steamships and railroads made gathering botanists from all over the world more feasible. The first congress, held in Paris, was an opportunity for many who knew one another's research from publications to finally have face-to-face discussions. Major issues included the ever-increasing plant diversity being discovered worldwide and the confusion of synonymy, more than one name for the same species. De Candolle's proposed code was not based solely on his own ideas. He attempted to describe a consensus view that could be accepted by the entire botanical community, though this consensus did not build rapidly. He proposed that the name given by the first person to publish a description of a species should be kept, unless there was a need to move the plant to another taxon, a general name for any level of taxonomic classification. Later it was decided that if the species description was altered, the earlier author's name would be given in parentheses, followed by the revision's author. In other words, history mattered.[13]

Specimens also mattered. Species descriptions were often based on herbarium specimens, along with illustrations, notes, and live plants when available. All John Torrey had in describing *Darlingtonia californica* were specimens gathered three thousand miles away. Though many specimens of *Darlingtonia* have been made since that time, the specimen Torrey looked at is still important because he employed it in enumerating the plant's traits, studying its structures under a microscope, and rehydrating the flower by soaking it in warm water to restore some of its dimensionality and form. The specimen is now at the New York Botanical Garden herbarium along with the rest of Torrey's collection (see fig 9.3a).

There are several other specimens that Torrey used in describing this species: some are also at NYBG, and one is at the United States National Herbarium in the Smithsonian, the repository for much of the Wilkes Expedition material, including the first *Darlingtonia* specimen. When the type specimen concept was proposed in the nineteenth century, most botanists saw it as applying to any material used in writing the original description. Over time some botanists, including Nathaniel Britton and others in the United States, argued that only one specimen should be designated as the type. They contended that if there were more than one type, all should be examined when studying the species, yet these might be distributed over several institutions, as with *Darlingtonia,* and since the preponderance of types were in Europe, this put a particular burden on non-European botanists. The difference of opinion led to a "schism" among taxonomists that was not resolved until the 1930 International Botanical Congress in Cambridge, England, where a new code was ratified. It stated that a single specimen, the holotype, would be designated for each species when its description was first published. There are other kinds of types, for example, paratype for other specimens used in the original description.[14]

Since the type specimen concept was finally codified in the twentieth century, and introduced only in the latter part of the nineteenth, what about species described by Linnaeus and others earlier in botanical history, such as the Lewis and Clark specimens (see fig. I.1)? After types became the touchstone of botany, there were efforts to designate types, termed lectotypes, for

these species. In many cases it is possible to label a specimen as a lectotype
if it was in Linnaeus's herbarium, which is now at the Linnean Society in
London, *and* there is evidence that he possessed it at the time he wrote the
description. In some cases, he never saw the plant but described it from an
illustration, as is true of a number of plants in Hans Sloane's publication on
Jamaica. Using illustrations as lectotypes for flowering plants is no longer
common, but there are many cases of this from the past, underlining why
accurate illustrations were vital to the development of botany and closely
allied conceptually to specimens.[15]

Because type specimens are important, they are treated with particular care in herbaria, kept in colored folders, often orange or red, so they can
be easily spotted. In many cases all the types in an herbarium are stored
separately, as were those at the California Academy of Sciences in San Francisco. Loss of many type specimens from the herbarium of the Berlin-
Dahlem Botanical Garden when it was bombed during World War II was a
major taxonomic catastrophe. In 1929, James Macbride, a botanist from the
Field Museum in Chicago, photographed thousands of type specimens in
the Berlin herbarium, mostly of South American plants. He wanted a reference collection to bring home for his research on the flora of Peru. These
photos have become an invaluable record of what was lost. This presages
the twenty-first-century effort to digitize type specimens and make them
available online. The project would make earlier botanists drool since types
are now so easy to access. In the past, lengthy searches through the literature were necessary to locate the herbarium that held a type, and then there
was often another search at the herbarium in question, since types were not
always identified as such in collections.

One last note on types: they are not ideals. In other contexts the word
type is used to designate an exemplar: the best or most typical individual.[16]
This is not the case in botany, though it can be for botanical illustrations, for
which artist and botanist may work to create an image that shows all the
qualities of a species to its best advantage. A type specimen just happens to
be the plant used in describing it by the first person to publish on it. There
may be better specimens of *Darlingtonia* than the one Torrey used, but if
he did not study them, none could be a holotype.

There is another term used to describe some herbarium specimens: *voucher*. It refers to a plant used in a particular study. It is physical evidence of what the botanist examined in describing its habitat, extracting DNA from it, or measuring characteristics like flower or fruit size. A botanical artist might deposit a voucher for a plant used as a model for an illustration so future researchers can check its accuracy. Botany is a discipline in which it is relatively easy to create a voucher, to archive its essential study material: the plant itself. It is more difficult and costly to create zoological vouchers often involving taxidermy. The practice is particularly important in botany due to the large number of plant species, many with very similar traits. Vouchers may not be considered quite as valuable as type specimens, but they are nonetheless essential to the process of doing botany in many of its applications. Studies not backed by vouchers are not as convincing because there is no physical evidence of precisely what species was analyzed.[17]

Though vouchers have been used for years, they are now requested more explicitly in order for articles to be accepted for publication, including those in systematics. This has not always been the case, and without reference to a specimen, it is impossible to precisely replicate the research. Also, botanic gardens have long made specimens from plants in their living collections, and this is becoming more routine as a way to document exactly what was growing in a garden at a particular time. Less common was landscape designer Beatrix Farrand's creation of two thousand specimens of species growing in her Maine garden; this was a teaching tool for her students. The sheets are useful records about a garden that no longer exists because each includes a map marked with the site where the plant grew.[18]

The Work of Botanists

After being collected, all specimens, whether vouchers or not, are sorted, have their identifications verified, and are then mounted and labeled. Specimens can be used in a variety of ways, including what's termed alpha taxonomy: identifying plants and finding new species, then describing and classifying them, along with classification revisions of already described species, genera, and families. The last step in this process, as in all science, is publication. No species officially exists until its description is published, and there are now explicit rules in the botanical code for what constitutes publication. The code is considered binding in naming plants and is revised every six years at an international botanical congress.[19]

Sometimes the steps from collection to description and publication do not follow one another in a timely fashion. An unknown may be found and collected, then stored in an herbarium for years as an unmounted specimen, perhaps with an identification that will later turn out to be incorrect, just a family designation, or even simply a question mark. Every herbarium has folders marked "indet.," short for the Latin *indeterminabilis,* meaning that the identity hasn't been determined. These await study by an expert in that family; some may eventually be identified as new species. A few years ago, a group of botanists studied publications on new species and found that only 16 percent were described within five years of collection and 25 percent waited fifty or more years. This has been a perennial problem in botany; a bellflower brought to Italy from Crete in the sixteenth century was not described until more than a hundred years later.[20]

A collector working in the institution housing the specimen might be able to fast-track a specimen's mounting, and the same may occur if it is sent for identification to an expert in that family or genus at another institution. Still, many herbaria have massive backlogs of specimens waiting to be mounted and have their identification verified. In the herbarium at the National Museum of Natural History in Paris, about eight hundred thousand unmounted and uncatalogued specimens were sorted out during a renovation. A survey of German herbaria revealed that over 20 percent of their specimens remain unmounted. There are many reasons for this, including the disruption of collections due to war, civil unrest, and political upheaval—or just plain lack of resources.[21] Even the heavy, acid-free mounting paper is a significant expense. That's why most duplicates traded among herbaria are sent unmounted, or considered worth more in exchange if mounted. Despite issues of space and money, some herbaria do not want to refuse collections that are offered in case there are some gems among the specimens, and often there are.

After being recognized as new, a single species may be described in a journal article, though frequently such publications deal with several species, and perhaps an entire genus or even a family, if it is not too large, with the new plant triggering a reexamination of its nearest relatives. There are also two book-length outputs of taxonomic research: monographs and floras. Monographs focus on a particular group of plants, often a genus or family. In the case of a very large family like orchids, the work may be published in several volumes by a number of authors, like Oakes Ames's orchid monographs.

The same can be true of floras, which deal with the plants of a particular city, county, or park, but can also cover more extensive geographic areas. The *Flora of China* was completed in 2012 with twenty-two volumes each of text and illustrations. The *Flora of North America North of Mexico* is a massive undertaking still in progress. After the *Flora of Tropical East Africa* was finished in 2012, biographies of its twenty-seven hundred collectors were compiled. They span more than two centuries, with vast differences in backgrounds and varied approaches to collecting, from utilitarian and aesthetic to scientific and environmental. The biographies reveal how over time, a botanical infrastructure is being built by Africans to continue the study of endemic or native biodiversity. The flora is a part of that infrastructure, making it easier to identify newly collected plants as novel species.[22]

Another impressive work is *Intermountain Flora: Vascular Plants of the Intermountain West, U.S.A.* in seven volumes.[23] A young botanist teaching in Utah, Bassett Maguire, conceived the project in the early 1930s. Trained in the eastern United States, he was fascinated by the very different plants he encountered in the West. Maguire proposed a flora on the region of the Great Plains and the Columbia Basin: the Intermountain West. Before moving to NYBG in the early 1940s, he amassed twenty thousand specimens, a solid start for the project. He continued to collect in the West during summers until the mid-1950s, when he shifted to working in South America and turned over the intermountain project to his former student Arthur Cronquist. Head of the venture until his death in 1992, Cronquist was supported by a team of botanists, with one or more focusing on each plant family. As botany became more professionalized, it also became more communal in order to cover the breadth of plant diversity.

While collecting is a major part of constructing a flora, the specimens then have to be identified, classified, and described. Writing the intermountain flora began in 1968, with the first volume appearing in 1972. Noel Holmgren, who began as Cronquist's student, played a major role in the project, not only through his taxonomic and editing work but in preparation of the illustrations. His art background served him well in laying out plates, that is, arranging multiple pen and ink illustrations on a page in an understandable and attractive format. It also aided his collaboration with illustrators, most notably Bobbi Angell, who produced over twelve hundred drawings for the flora.[24]

The last volume of the intermountain flora appeared in 2017, almost ninety years after the idea for the flora was conceived and forty-five years after the publication of the first volume. This is not an unusual time span for a large project; it is a massive task to deal with dozens of authors and thousands of specimens distributed over many institutions. The last volume includes a key to the families in the flora. A key is a vital part of a taxonomic work, a guide on how to identify a plant. Many keys are dichotomous, meaning they present a given attribute, such as leaves, with two descriptions, for example, simple or compound. Depending on which is chosen, the user is then led to the next attribute duet until a species determination is made. It is not easy to construct keys, and reference to herbarium specimens is essential. It isn't easy to use keys either, and learning the art is part of every taxonomist's training. Another helpful taxonomic tool is a checklist, often an early step in the development of a flora. It's a list of the species in an area, valuable for professionals and amateurs in learning about the plants of a park, city, or county, for example. Lists date back to the early days of botany: Ghini and Aldrovandi worked on a list of plants in the Pisa garden, and Linnaeus's guide to George Clifford's garden was an elaborate version of such a list.

Field guides fall somewhere between lists and floras and take a variety of forms, many with photographs or drawings, some encyclopedic, others less daunting, but all usually in a relatively small format so they travel well. Some are aimed at professional botanists, others at amateurs with various levels of sophistication. For example, arranging a guide to wildflowers by flower color in not considered very useful to a botanist, who is looking for classification, but it is a godsend for a novice trying to identify a roadside plant. No matter what the level of sophistication, all guides are grounded ultimately in taxonomy and in specimens.

This tour of systematics is a prelude to the next chapter, which explains how it fits into the larger picture of botanical research more generally. Though work on plant growth and development had been going on for centuries, as had study of the relationship between plants and environment, these studies took on new meaning as botanists investigated the many implications of Darwin's theory of evolution.

Changing Botany

AS THE EXAMPLES OF THE *Flora of China* and the *Flora of North America* suggest, taxonomists worked actively throughout the twentieth century organizing plant information and describing new species. This was especially true in the species-rich tropics. Some particularly energetic collectors, such as Thomas Croat of the Missouri Botanical Garden, managed to amass an enormous number of specimens in a long career. It is easy for most collectors to remember precisely how many specimens they have gathered because it's standard practice to number a botanist's collections consecutively, starting with 1. Few are in Croat's league of over one hundred thousand specimens, and some even disparage such work, calling it "baling hay," since the result is a pile of dried plants.[1]

This attitude became more common as the twentieth century progressed and other fields of biology moved to the fore. It was not so much that taxonomy was abandoned as a specialty, but that it had to compete with other disciplines within botany for attention. Some research areas had been developing for centuries. Study of plant structure began in the seventeenth century with the work of plant anatomists like Nehemiah Grew, who used a microscope to examine the interior of stems and leaves and began to puzzle out how they were constructed. Others were interested in plant chemistry and physiology. The eighteenth century saw the discovery that plants absorb oxygen and release carbon dioxide into the atmosphere, an early step in understanding photosynthesis, how plants use the sun's energy to make sugar.[2] At the same time, agriculturalists and horticulturalists investigated optimal growth conditions for cultivated species, incorporating research on physiology and development.

Ecology

These disciplines became more clearly delineated by the end of the nineteenth century and more developed in the twentieth with the addition of two new fields of interest: ecology and genetics. Here again, there were antecedents. Ecology grew out of the writings of Alexander von Humboldt on biogeography and observations by generations of plant collectors who noted the different conditions under which plants with particular characteristics were likely to be found: leafless, thick-stemmed cacti in deserts, diminutive species at high altitudes, and carnivorous ones in nitrogen-poor bogs.

Many early ecologists like Charles Bessey began their careers as taxonomists. He moved into survey work on Nebraska grasslands and taught students to systematically sample so they could quantify their results. He was influenced by German botanists who took the lead in nineteenth-century laboratory-based research on cellular anatomy, the relationship between plant structure and function, and how these were influenced by adaptation to different environments. They took advantage of German colonial ambitions to do research in botanical gardens in South America and Asia, collecting with an eye toward studying the effect of environment on plant growth. Influenced by evolutionary theory, German botanists investigated how plant communities changed over time, which became an important aspect of ecology.[3]

One of Bessey's students, Frederic Clements, studied change in a specific type of plant habitat, the forest. He advanced the concept of succession: fast-growing tree species are successful early in a forest's development, but eventually slower-growing trees may become tall enough to overshadow the first growth, leading to its demise. In time, an equilibrium or climax forest would appear and remain relatively unchanged. Not everyone agreed with this analysis, and as ecology developed, relationships among plants and animals were studied in greater detail and became more nuanced.[4] This research did not necessarily involve close attention to taxonomy. It was often more about studying food webs and measuring how energy flowed from plants to the herbivores that ate them, and then on to carnivores that preyed on herbivores. Species identification was often neglected. In some studies identification was still important but seen merely as a service to ecological research. Ecology became more "glamorous" than taxonomy, especially from the late 1960s on when the environmental movement gained momentum.

In the mid-twentieth century, degradation of air, water, and land was a focus of attention in the developed world, though such deterioration—and concern about it—was centuries old. Even in the Middle Ages there was discussion about where cutting down forests would lead, and Mark Catesby noted how agriculture in the colonial Carolinas led to destruction of habitats for many animals and plants.[5] But while the health of the land had long been compromised, the realization that this damage affected human health finally led to significant action. In the 1950s, Britain passed laws to improve air quality in large cities and prevent future killer smogs like London's in 1952. In the United States, the creation of the Environmental Protection Agency as well as the passage of the Clean Air Act in 1970, with other regulations following, brought a slow but steady improvement in environmental conditions in many areas.

Enforcing these laws required monitoring air, water, and land. Government and private moneys poured into these efforts, including creation of environmental impact statements before changes in land use were permitted. These sometimes necessitated surveys of plant and animal life in an area. At the very least, consultants working on such projects contacted herbaria and taxonomists for help in identifying the species they encountered. In some cases, specimens were collected to record what was found, but the emphasis was on living nature out in the field, not dead plants in cabinets.

Genetics

Genetics, even more than ecology, is a twentieth-century science. It began with work by plant breeders who substantiated Gregor Mendel's rediscovered publications from the 1860s on pea plant crosses. Mendel himself documented his research in specimens still kept at the monastery where he grew his plants. When rediscovered, the significance of his work was viewed in the light of Charles Darwin's theory of evolution, which made the search for a plausible explanation of how traits were inherited central to the work of biology. However, interest in heredity stretched back to ancient times. As Darwin pointed out, plant and animal breeders routinely selected individuals with favorable traits for propagation and had done so throughout the history of agriculture. This led to more sophisticated experiments, such as Thomas Fairchild's "mule." Grown about 1717, it was the first successful hybrid between two plant species.[6] By the time Mendel's studies were unearthed around 1900, cell biologists had begun to link heredity to structures in the cell's nucleus, chromosomes, with each species having a particular number. During cell division, chromosomes sort themselves out so each new cell retains this number, the chromosomes having been duplicated before the cells divided.

Further study corroborated the importance of chromosomes: the botanist Cyril Darlington discovered their behavior during meiosis, the process by which sex cells are formed. Darlington worked at the John Innes Horticulture Institution, a British botanical research center with a long history of investigations in agriculture and systematics. It housed an herbarium, which was of little interest to Darlington, who focused on the cellular level. When he became director at Innes, he had to visit the collection from time to time, and its curator derived satisfaction from the fact that the very tall Darlington usually cracked his head passing through the very low entryway.[7] Taxonomists and cell biologists definitely belonged to different botanical tribes.

By the 1950s, botanists were able to count chromosome numbers accurately, adding another data type to the study of species traits. Catalogues of chromosome numbers for plants were constructed, and it became clear that a species might have double or triple the counts of related species due to unusual occurrences during cell division. This condition, called polyploidy, is common in plants, with analyses indicating that many species probably arose through such increases in chromosome number in the past. Because of interest in this issue, it is not uncommon to find an herbarium specimen with a notation that its tissue was used in chromosome research.[8]

Chromosomes seemed important in transmitting traits from one generation to the next, yet they contained several chemical constituents, so the question became: were all the substances involved or just one? The 1940s brought the first evidence that one type of molecule—DNA—transmitted genetic information. This finding became more convincing ten years later when DNA's linear, two-stranded structure was discovered. This suggested how DNA could both carry information in the order of its building blocks within each strand and replicate by separating and copying its strands.[9]

After the discovery of DNA's structure, researchers worked for another decade to determine how sequences of the four building blocks, or nucleotides, in DNA contained information to guide a cell's and an organism's structure and function. To put it simply, a segment of the DNA sequence—a gene—contains the code in the linear nucleotide sequence to make proteins, the molecules responsible for the structure and function of cells. An organism's total DNA, its genome, may have billions of nucleotides, with a single chromosome having millions of them. More decades were needed to find ways to break down these long strings into shorter segments that could then be further broken down, each unit identified, and the segment thus sequenced. Billions of dollars and ten years of research resulted in sequencing the human genome, which obviously received public attention, but many species of plants, animals, bacteria, and viruses were also investigated.[10]

One reason for interest in DNA sequences and in making comparisons between them is that, as Darwin long ago suspected, there is a strong relationship between genes, that is, DNA sequences, and an organism's traits. Each species has a distinct order of nucleotides, its genome. Usually, the more similar the arrangement in two species, the more closely they are related. However, results can often be interpreted in more than one way. For example, what does "similarity" even mean? What percentage of the sequence has to match? Are all types of matches of equal significance?

Despite these issues, taxonomists are incorporating genetic analyses into their work on phylogenetics, that is, the study of evolutionary relationships among plant species. It is so central that the term *systematist,* one who studies relationships, and *taxonomist,* one who describes species, are almost interchangeable. Botanists do not all value different types of data equally, leading to disagreements in systematics. For some, sequences are of prime importance: the DNA evidence decides the case if it indicates a close connection between two species, even where external traits suggest otherwise. Some are less convinced of molecular data's preeminence. They argue that much is unknown about environmental influences on how genes function; a broader set of characteristics should be considered, including the floral traits used for centuries in classifying plants, as well as ecological data.[11]

Those outside science often consider scientists cool-headed, rational beings who make decisions based on the unemotional study of evidence. This is an ideal, but there are two problems with the picture. Evidence is not always clear-cut; there are often data for and against a particular position. Also, scientists are human beings. Because their work is important to them, they have strong emotional as well as intellectual ties to it. So arguments in systematics continue, and herbarium specimens are important in these debates for studying differences in plant traits across species. Today, herbarium specimens have become even more important because they can be sources of DNA for sequencing, though there are difficulties isolating DNA in some plant groups and preservation methods.

Over time, sequencing costs decreased and techniques became more sensitive, so the partially degraded DNA in herbarium specimens could be accurately analyzed. Methods are now precise enough to sequence DNA from centuries-old plants, including that of a tomato plant sequestered in an herbarium, the work of one of Luca Ghini's colleagues (see fig. 2.1). In addition, managing the data produced in this work has become more sophisticated. The only way vast strings of sequences can be understood is through computer storage and analysis. Computer hardware and software are essential to what is called bioinformatics, the field of creating and analyzing large biological data sets.[12] The sequences are deposited into massive databases. When a new gene is sequenced, the results are fed into a computer database that is searched for similar sequences. Analysis may result in matches or near matches rated for degree of similarity. If it is below a certain threshold, systematists may consider moving a species to another genus, even another family, both triggering a name change. The second name of the binomial,

the species epithet, will usually remain the same while the first part, the genus name, will change.

Revisions in Systematics

By the latter part of the twentieth century, large-scale revisions in plant systematics were underway, thanks in part to DNA sequencing results. Fungi
used to be considered plants that had no chlorophyll and therefore could
not photosynthesize. Then ecological, physiological, and genetic data revealed that they were not closely related to plants, and investigations into
their structure and physiology substantiated this. So fungi were removed
from the plant kingdom. The same thing happened to many aquatic "plants,"
such as the red and brown algae, which are not related closely to plants either.[13] Since a lichen is composed of an alga or a photosynthetic bacterium
living as one with a fungus, this group was also reclassified, based on the
fungal component. The reexamination posed a problem for botanists, some
of whom specialized in studying these groups: were they then ousted from
the discipline, and were specimens exiled as well? No field wants to lose
ground, or specimens, so in most cases the experts and the collections stayed
put. Many herbaria have extensive collections of fungi, lichens, and algae.

DNA data have also led botanists to reexamine plant genera and families. As a result of sequencing, some plants were put into different taxa and
many names changed. Herbarium sheets reflect such shifts, with small
pieces of paper called determination slips attached (see fig. I.1). Old labels
are not removed, to retain a record of what the plant used to be called, helpful in referencing older literature. The guide for this work is the *International Code of Botanical Nomenclature for Algae, Fungi, and Plants,* which
provides rules for naming and renaming the three groups. The code is ad-
ministered by a committee of systematists who interpret and enforce the
rules. For flowering plants, angiosperms, its decisions are based on the
work of the Angiosperm Phylogeny Group (APG), whose latest report is
APG IV (2016). The group has assigned genera to families based on evolutionary relationships among species. Since DNA sequencing is ongoing,
this work has to be reexamined along with other taxonomic data. APG IV

was published seven years after APG III, and no systematist considers it the final word.[14]

This leaves herbarium administrators with a problem: how to arrange specimens in a collection and when to rearrange. These questions have no clear answers, so some larger collections still rely on systems that were put in place years ago. New York Botanical Garden uses the arrangement of families that Arthur Cronquist published in the 1980s. The Royal Botanic Gardens, Kew, recently moved to the APG III system, as did the Paris herbarium when it reshelved the entire collection after a massive renovation. If these institutions want to align with APG IV, it will require some shifts, but nothing too extreme.

What does this realignment mean for particular groups of organisms? One of the stickier cases is that of *Acacia*, a large genus native to Australia, Africa, Asia, and the Americas, though there are more species in Australia than elsewhere. Recent genetic evidence indicated that these plants should be classified into three different genera, with the 960 Australian species in a genus of their own. There was little disagreement about this; the argument was about what these genera should be called. According to the botanical code, the genus name *Acacia* should remain with the African species. Why? Because it was to an African species that this name was originally given, and the rule in the code is that naming is based on priority. Consequently, the other species should be put into other genera, with the name *Racosperma* suggested for the Australian species and the remainder called *Senegalia*.

Renaming a relatively small number of plants *Senegalia* was not a big problem; such shifts have occurred many times in botany. However, *Racosperma* did not sit well with Australians, in part because it lacked euphony, though their objections extended beyond that. Plants are important to Australians, as are animals. Australia has been separated from other continents for so long that its isolation resulted in a wonderful diversity of endemic animals, like the emu and platypus, and plants, such as *Banksia*, some *Eucalyptus*, and many *Acacia* species, including Australia's national flower *Acacia pycnantha*, the golden wattle. This plant was at the heart of the *Acacia* nomenclature issue, though it did not provide the principal taxonomic argument for keeping the *Acacia* name with the Australian species.

The justification was that there are more Australian acacias than any others, and therefore it would lead to more nomenclatural disruption to rename them all. This was considered a valid argument and has been used before in other disputes, though usually the disruptions entailed fewer species and were without news media involvement.

The naming question became a cause célèbre in Australia, though there was also substantial attention paid to the issue in Africa. *Acacia karroo*, the sweet-thorn tree, is a significant symbol in South Africa and throughout the continent since these trees are striking forms in savanna landscapes. The dispute was ultimately resolved by the committee administering the botanical code at its meeting before the 2011 Botanical Congress, which happened to be in Melbourne, Australia. The committee recommended that the *Acacia* name remain with the Australian species, with the genus including the African thorn tree to be called *Vachellia* and three other genera created for the remainder. The congress accepted this solution, which was met with distaste by many, who saw it as yet another example of dominance by rich nations with scientific clout over the less wealthy with little scientific influence.[15] In any case, the story highlights the capacity of taxonomic nomenclature to ignite passions in botany and beyond.

Fading Herbaria

While botanical systematics was undergoing dynamic change, the biological community as a whole was moving in other directions, and this presented a problem for herbaria with growing collections. Institutional space is often allotted on the basis of status, effectiveness, or economics. A lab bringing in large grants and/or publishing in prestigious journals will gain square feet in order to generate more results. After World War II, the United States sought to maintain the scientific apparatus developed during the war by creating funding institutions, particularly the National Institutes of Health for medical research and the National Science Foundation for basic research. Since cellular biology, genetics, and ecology were generating more attention and attracting more students to their ranks, they became the areas of biology receiving the most funding—and space. More traditional fields such as zoology and botany, the organismal disciplines, lost ground.

At midcentury, most herbaria that were staffed with active taxonomists were growing as new specimens were added; this meant they eventually ran out of space. In the past, this would mean expansion. Now such requests were met with resistance: why fund a discipline that generated little or no revenue and did not publish articles in "hot" journals? Molecular biologists investigating genes and proteins were clamoring for laboratory space, and herbaria often represented large territories ripe for annexation. Also, when hiring questions arose, departments tended to ignore systematists or organismal biologists, so the need for herbaria dwindled.

Ultimately, these trends led to decreased collecting. One study found that specimen acquisitions in U.S. herbaria have been declining since the 1930s. There are exceptions, with impressive growth by some, but the majority have seen decreases. Lack of space is one deterrent but fewer faculty and fewer botany courses mean there is less collecting, by students as well as faculty. Another issue is that many collect the genus or family they are studying, ignoring other plants that may be in the area. Broad surveys are less common. This phenomenon is not limited to the United States; it is occurring worldwide, with some institutions discouraging collections outside areas of interest.[16] This will make longitudinal studies of changing habitats more difficult in the future.

With such problems, it's not surprising that some formerly active herbaria faded or dissolved. Others became inactive, meaning that there was no curator or manager to lend specimens to researchers in other institutions who might need to consult them, nor was the collection augmented with new sheets. Cabinets sat untouched, and there are still collections in this limbo. If and when they are finally inspected, discoveries could be grim; lack of attention can mean insects invaded, chewing their favorite plants to shreds, or water damage from leaky pipes or roofs caused devastation, problems especially likely if the collection was moved into offsite storage to free up space.

Sometimes, unwanted herbaria were disposed of. As one commentator wrote: "Some years ago, personnel at a well-known university literally defenestrated its natural history collection, tossing them out the window; then the administration had second thoughts and put it in storage."[17] There were other instances of destruction, near destruction, or, at the very least, selective destruction of what at the time was seen as not worth keeping. The most common outcome for unwanted specimens was to ship them to an institution that was willing to take them. Usually, it was a larger herbarium absorbing a smaller one. NYBG was known in the 1980s for acquiring "orphan" collections that often contained valuable specimens. DePauw University's herbarium had many types specimens, and Wabash College had a good number of nineteenth-century sheets from noted collectors of the time.

An item in the Wabash collection suggests the hidden treasures found in herbaria considered expendable. One sheet contains three specimens of the cutleaf ironplant (fig. 12.1). The two specimens on the right were collected by Aris Berkley Donaldson in 1874 while on General George Custer's expedition into the Black Hills of South Dakota, two years before the battle of Wounded Knee. Donaldson sent the plants to botanist John Merle Coulter for identification. Then at Hanover College, Coulter moved to Wabash College in 1879, taking his herbarium with him, despite Hanover's attempt to purchase it. Coulter later added a third specimen to the sheet, one found by the energetic California collectors John and Sara Lemmon in the San Bernardino Mountains in 1880. The Lemmons were friends of Coulter's and advertised their collections in the journal he edited, where he even ran a notice of their marriage. So this one sheet relates not only to botany but to history, government, and botanical economics. It was only when the specimen was examined at NYBG that its significance was discovered—Coulter had written on the label, "Custer's Exped. Black Hills"; thirty-nine sheets were so marked.[18]

The Wabash herbarium was rather small, and therefore relatively easy to absorb by a large institution like NYBG, but there was at least one case of the reverse. In 1976, Stanford University's Dudley Herbarium contained 850,000 specimens and was assimilated into the CAS's collection of 600,000 specimens. The merger was years in the making. The impetus for it first surfaced in the 1950s when, to increase the university's academic status, Stanford administrators decided to focus on "steeples of excellence,"

Figure 12.1. Specimens of cutleaf ironplant (*Aplopappus spinulosus,* now *Xanthisma spinulosum*). The two on the right were collected by A. B. Donaldson, and the one on the left by John Lemmon and Sara Lemmon. The circle in the upper left pictures the fragments in the packet above it. Packets are now often added to sheets when material has fallen from the specimen. (Image courtesy of the C. V. Starr Virtual Herbarium [http://sweetgum.nybg.org/science/vh/])

research areas of national importance and therefore attractive for government funding. Systematics was definitely not the steeple it had been earlier.
There was resistance from those who had built and used the collection, but
as years went by this group dwindled by attrition. Eventually the National
Science Foundation funded a state-of-the-art facility at CAS to accommodate both collections and allow for further expansion.[19]

The lack of an herbarium at Stanford troubled some. Stanford botanist John Hunter Thomas attempted to continue work in systematics by
dividing his time between Stanford and CAS. David Ackerly, who joined
the faculty in the 1990s and did research on plant evolution that involved
DNA sequencing, ultimately moved to the University of California, Berkeley, which has a large herbarium. He needed to reference the plants themselves and found his molecular systematics work more vibrant at an
institution with a large plant collection.[20] He realized it would be difficult
for Stanford to create an adequate collection because it would have to start
from scratch; it would be impossible to build a collection equivalent to one
that spanned well over a century of collection. Despite this, there is now an
active herbarium at Stanford, housed at its Jasper Ridge Biological Preserve. It is not a large research collection like Ackerly needed, but a teaching collection with mostly native California species to give environmental
science students knowledge of what grows in the state and what needs to be
conserved.

Transfer of specimens to institutions that can properly care for them
is definitely a good thing. Curation, that is, management of a collection, is
essential. This means acquiring new specimens through active collection
programs, keeping sheets updated with name changes, receiving gifts from
collectors, trading specimens with other herbaria, and allowing specimens
to be borrowed. It has been traditional among herbaria to freely lend materials to researchers in other institutions, though loans can present problems.
Packaging dozens if not hundreds of specimens is time consuming and
costly at both ends, and there is the threat of damage or loss. Curation also
involves exchange. Botanists often collect plants in multiples, except when
species are rare. Called duplicates, these can be traded with botanists in
other geographic areas. The recipients usually send specimens in return, so

both collections benefit from the trade. Another curatorial task is sending out collected material that cannot be definitively identified to an expert in that family. The specimen is customarily given as "payment" for the service.

Sharing specimens requires much bookkeeping: logging loans and exchanges. When loans are returned there are usually new determination slips, either changing a name or verifying it. In the process of filing and refiling specimens, curators discover much about their collections: items that may not have been properly identified or were misfiled, insect damage, loose plant material that needs to be reglued to a sheet. One curator had an inquiry about a single specimen, and in the process of answering it, found thirteen new type specimens and one from Linnaeus, all now with updated labels.[21] Managing an herbarium is like managing a household: the work is never done and neglect leads to bigger problems.

Though specimens may find good homes in other collections, institutions that cede their herbaria are poorer for it. If they are educational institutions, then students no longer have access to a fundamental resource for their biological education: the plants themselves. Herbaria personnel are frequently the people who identify plants; Rupert Barneby, a noted botanist at NYBG, estimated that he made about four thousand identifications in a year, and that each took about ten minutes to ID.[22] A nonbotanist is more likely to reach out to a local institution for an ID and not bother if it is not easily accessible.

But the loss of an herbarium, or diminishment in collecting, is usually not widely mourned outside the botanical community, and this is partly the fault of herbaria. Botanists were not good at advertising their resources, and in some cases did not want to. They saw herbaria as places of research, not geared to public education. This attitude is changing now, at least in part due to the realization that the survival of a collection depends on how much it is valued and how many different groups value it.

THIRTEEN

Useful Plants and Ethnobotany

𝒦˜

IT'S AN UNDERSTATEMENT TO SAY that plants are endlessly useful to humans. They are beyond useful: responsible for life on Earth as we know it, converting carbon dioxide and water to sugar and oxygen, both fundamental to the planet's energy flow. Plants provide food, medicine, and materials to clothe, shelter, and make life richer with everything from chewing gum to violins. Interest in plants has always centered on how they can benefit humans, with early modern botanists focusing particularly on medicine. When Luca Ghini was creating his herbarium in the 1540s, he was teaching materia medica, medicinal botany, a field created by the ancient Greeks and Romans. But Ghini and his fellow botanists were slowly widening their range of interest beyond medicine to observing an array of species, including those providing food for the kitchen and beauty for the garden. In early herbaria there are plants collected simply because they were novel or had fascinating characteristics, but the preponderance of specimens were from useful plants. This trend has continued, with such plants overrepresented in herbaria to this day.[1]

Despite the broadening of focus, medicinal botany remained a major driver of plant collection and botany an essential part of a physician's education into the nineteenth century. Pharmacists, while also deriving medicines from animal and mineral materials, dealt heavily in plants and used them in many of their compounds. Both pharmacists and physicians often created herbaria for reference, to guide students, and to keep track of new species that might provide better remedies. Others were also involved in studying, gathering, and using plants for healing. Most religious institutions had herbalists among their members to furnish remedies in-house, and many women grew useful plants in their gardens, shared them with others, and were often called upon for medical advice.[2] Such practitioners had vast stores of knowledge, but even the literate among them might merely jot down notes in the margins of herbals. Their experiences were rarely shared more widely and in some cases were belittled and even suppressed by more educated medical practitioners who disliked competition but were not above learning from those in lower ranks. As in all fields, there was a range of relationships among these different classes and slowly information trickled up and down social strata.

Similar tensions, though often more intense, arose when explorers began to learn about medicinal botany in other parts of the world. There were added layers of complexity because travelers and residents shared neither language nor customs, nor in many cases ailments, at least early in their exchanges. Since most explorers had relatively short stays in any one area, they rarely developed a deep understanding of the flora or of indigenous plant use. There were exceptions; missionaries often remained for long periods and learned local languages. They sometimes collected specimens to send home for identification and recorded the local names for plants. These vernacular names were important for later botanists who revisited areas seeking similar species.

Colonization is rightly seen as subjugation of indigenous peoples and exploitation of their resources. What may be less obvious is that these resources encompassed a great deal of indigenous knowledge, including about native plants and their habitats. For a European to enter a Mexican desert, mountainous regions of India, or African rainforests was to encounter strangely foreign flora. Making sense of the new environment, especially with barriers of language and custom, was difficult. Almost every plant explorer relied on native guides, both to deal with travel through difficult terrain and to locate plants, name them, and explain their properties and uses as food, medicine, tools, and so on.[3] This reliance had begun with the first voyagers, but it became more systematic over time. Often, guides' names were left unrecorded and written accounts of new plants rarely detailed how they were discovered, so it is easy to credit explorers like Humboldt and Bonpland with finding plants that they might easily have overlooked. Humboldt's comment about collecting only a tenth of what they saw hides the fact that how they chose was often dependent on the eyes and knowledge of others.

Economic Botany

Some plant hunters were interested in plants for a wide range of uses beyond medicine, including textiles, foods, timber, and horticulture. By the nineteenth century, the term *economic botany* described this broad curiosity about any plants with monetary value. Global powers took what they wished from the lands they conquered, exerting not only political but economic control, extracting mineral and biological wealth. For millennia, the remote Maluku Islands in Southeast Asia were the sole source of cloves and nutmeg, but the indigenous peoples there did not reap the rich rewards of a huge markup that meant profits for the Dutch controlling the islands' ports.[4] Eventually this lucrative market waned when cloves were grown by the French in Africa and nutmeg plants were spread to several areas of the British Empire. The same thing happened with relocations of cinchona, vanilla orchids, rubber trees, and a long list of other species.

Sometimes colonial plants took unexpected turns. Breadfruit was among the species Joseph Banks and Daniel Solander collected on the Cook voyage. Banks later considered it worth investigating because of its nutritious fruit: it could provide nourishment in the West Indies as a cheap food for the enslaved and indigenous sugar plantation workers. Banks planned a scheme to ship breadfruit from the South Pacific islands where it was endemic. William Bligh was put in charge of the expedition, which is famous because of its crew's mutiny, but that's not the end of the story. Bligh was sent out again and successfully transported live plants from Tahiti to St. Vincent and Jamaica, where they flourished. However, workers were not interested in eating this unfamiliar fruit, suggesting the strong cultural component in food choice and the limits to imperial manipulation. Later, breadfruit was accepted and is now an important commodity in the West Indies.[5]

With botanical resources being exploited and their riches profiting those in distant nations, it's no wonder that citizens of developing nations, and particularly indigenous peoples, are leery of foreigners collecting plants, making the term *economic botany* less popular today. It was used by colonial powers like Britain to highlight how they had discovered and appropriated many plants and their uses from indigenous peoples. As discussed earlier, economic botany collections, now often termed biocultural collections, were popular in the nineteenth and early twentieth centuries to let the Northern Hemisphere public know about wonderful products derived from plants and ingenious uses found for them, from arrow poisons to basketry to bark cloth (fig. 13.1). When these displays became less popular and literally gathered dust, most were dismantled. The Royal Botanic Gardens, Kew, had four economic botany museum buildings; now the display is reduced to a few beautiful old cabinets with remnants from the collection set amid tables in a café housed in a former museum building.[6]

The Kew economic botany collection still exists, kept in a storage facility and well curated, unlike items in many other such museums that were severely culled, sent elsewhere, or simply tossed out. The fluctuating value of collections wasn't often taken into account. Today, Kew's vast store, while no longer of economic importance, is a cultural treasure, with many items seen as works of art or as documents of indigenous peoples' lives in the past. Kew has worked with indigenous groups in Amazonia, sharing information about Richard Spruce's collections in the nineteenth century from their region and learning from them about how the same plant materials may or may not be valued today.[7] This effort is part of a larger one to assist in providing for biodiversity's future in the region.

Kew's work is also an example of the "decolonizing collections" movement: changing the way art and natural history museums regard their holdings and their mission. It means looking into the history of objects and asking difficult questions about how the items were acquired and for what uses they were appropriated, as well as exploring how cultures where the objects originated can be given greater recognition and how the items might be used today to enrich these cultures. Decolonizing a collection may involve taking note of derogatory language on labels, attempting to discover

Figure 13.1. Nineteenth-century lacebark fan made in Jamaica from the inner bark of *Lagetta lagetto*. The item is something of an herbarium, with fern fronds attached to the bark. This is an example of the plant products found in the Kew Economic Botany Collection. (© Royal Botanic Gardens, Kew. Photo: Joanne Muhammad)

the identities and histories of collectors and makers, and ferreting out links to slavery, forced labor, and other unsavory practices.[8]

In all cases, this work means enriching the stories attached to objects, adding to their value by illuminating the cultures from which they came. Herbarium curators are looking anew at items on their shelves, in part to attract new audiences. As discussed earlier, plant collecting was for many years tied to the slave trade, both in terms of where rich collectors' wealth originated and how slave ships gave plant hunters access to areas—and indigenous knowledge—that would otherwise have been unavailable to them.[9] But the past erasure of history goes deeper; often plant hunters failed to even acknowledge the assistance they received from indigenous collectors and support staff. Many at least noted local names and uses for plants, but these notes often disappeared from the specimens when later curators relabeled them with scientific names, considering their origins of no value.

At times, the information about the real activity of collection was written not on labels but in field notes or letters, so evidence may still be found by those willing to search for it. Erik Mueggler discovered much about George Forrest's work with collectors in China over thirty years by investigating archives of this twentieth-century Scottish plant collector. Forrest did acknowledge his chief assistant, Zhao Chengzhang. However, others doing fieldwork for him were anonymous, though they were expert in distinguishing between different species and varieties and in knowing where to search for rare plants or for ones with narrow ranges. When back at the base camp, Forrest and Zhao would examine the plants and sort them. After Forrest had written new labels, he would usually destroy the original notes in the collectors' languages, which Forrest did not understand and which Zhao translated. It is remarkable that the work of this team was so productive, considering the multiple language barriers; sign language and sketches proved useful, as did years of familiarity within the group.[10]

Ethnobotany

As the twentieth century progressed and economic botany went into decline, ethnobotany developed, incorporating some of the former's collections and concepts. Ethnobotany can be defined as the study of the relationships among plants and people—in other words, learning about plants through people's attitudes toward and uses of them. Because ethnobotany grew out of studies of plants and indigenous peoples, its emphasis has been on plant use among these populations and is often related to anthropological research.[11] Richard Schultes was a botanist instrumental in making the transition from economic botany to ethnobotany. He studied economic botany with Harvard's Oakes Ames in the 1930s and wrote his undergraduate thesis on the use of peyote cactus as a hallucinogen among the Kiowa people of Oklahoma. Schultes went on to pursue both systematics and ethnobotany. He collected specimens because he needed to be able to identify the species he found in the field and to study their relationships not only to human use but to one another.

Oakes Ames stressed the need to learn as much as possible about plants and their uses from indigenous peoples because these groups and their life styles were so vulnerable to diseases of civilization and to cultural disruptions. Schultes heeded his mentor's warning and recorded much about the groups he studied in tropical South America, especially their use of hallucinogenic plants in religious rituals. He wrote a great deal and also educated the next generation of ethnobotanists.[12] Ethnobotany is still evolving as a discipline and in the twenty-first century it is becoming more central to efforts promoting biodiversity conservation, equitable distribution of resources, and sustainability. However, botany's ties to economics and colonialism continue to influence these goals.

Though there is research on indigenous plant use in areas of North America and Europe where traditional uses of plants remain important, much ethnobotanic work is done in developing countries. Despite vast geographical expanses and travel challenges, richer biodiversity nearer the equator remains a magnet for collectors. Numerous U.S. institutions collect plants in Latin America and several in areas of Africa, particularly the extremely species-rich island of Madagascar, where French botanists also work because of France's long history there. German botanists continue to be active in East Africa and in parts of the Middle East.

The British collected throughout their empire during the first half of the twentieth century and still work in former colonies through institutions such as the Royal Botanic Gardens at Kew and Edinburgh. However, large-scale expeditions are long gone, and collecting in the developing world is now tightly regulated, one result of a series of international agreements coming out of United Nations–sponsored conferences. The Convention on International Trade in Endangered Species of Wild Fauna and Flora (CITES) was presented at a Washington, DC, meeting in 1973. It controls commerce in endangered species and comes into play in botany especially with illicit trade in timber from rare trees and exotic plants such as orchids and cacti. It was an early international agreement on the crucial need to preserve Earth's biodiversity.[13]

A meeting in Rio de Janeiro in 1993 led to the Convention on Biological Diversity (CBD), which gives each nation sovereignty over its biological wealth. The CBD aims to prevent developed nations from continuing to exploit the biota of developing nations, most having had their progress thwarted by centuries of colonial rule. The CBD represents a major shift in how the biological resources of a country are seen not only economically but politically and culturally. The cinchona tree, native to Peru, is more than the source of a valuable commodity, quinine. It also has a long cultural history; it was Peruvians who originally discovered its fever-relieving effects centuries ago, and this plant has been documented with herbarium specimens, seeds, and also in poetry and art. It is an integral part of Peruvian heritage.[14]

Cultural connections can be found for thousands of plants worldwide, but economic and political issues often are at the fore when it comes to plant collecting. Since the ratification of the CBD by most of the world's nations, these issues have had a significant effect on botany and on herbaria, just as they have had throughout the history of botany. The difference now is the aim of equitable distribution of value. Even nonsignatory nations must comply with the procedures set down by those that have signed if they want to be allowed to collect, so this and the other international agreements have had a significant impact on how the biological wealth of nations is viewed and treated.

This is especially true since ratification of another UN-sponsored document, the Nagoya Protocol, a 2011 agreement on access and benefit sharing that grew out of the CBD. It deals with the genetic resources of plants and animals, and how to ensure that their benefits are shared and used equitably. It aims to prevent exploitation—for example, by drug companies collecting plants that have medicinal uses and then developing drugs based on them in their home country's laboratories without sharing profits with the country where the plant was collected and with the people who revealed its medical efficacy.[15]

The Nagoya Protocol gives a host country the right to set strict limits on what can and cannot be collected, and also on its use afterward. A botanist who wishes to collect in another country must obtain a permit, or a series of permits, to do so. These delineate what can be gathered, often only specific plant groups and in specific quantities. Travel might also be limited to particular geographic areas. Where possible, unless the plant is very rare, it is collected in multiples, and specimens are retained in an institution in the host country as well as in the collector's institution.

Requirements and procedures vary widely from country to country, and it may take months or years to obtain necessary documents, which in some cases are issued across multiple government agencies at levels from countrywide to state, municipality, or other jurisdiction. This sounds daunting, and it can be. Some botanists and policy makers argue that the paperwork can be so difficult as to effectively prohibit or severely curtail collection and therefore hinder biodiversity research. So although the protocol does prevent exploitation, in certain instances it discourages research that might lead to practical benefits for the country in question. In a commentary on a recent assessment of biodiversity on the island of New Guinea, the authors noted that while half the island is part of Indonesia, a signatory to the protocol, the other half, Papua New Guinea, is not. This makes the latter a more attractive location for biodiversity research.[16]

Medicinal Botany Updated

There are good reasons why botanists continue to collect in the face of regulations. Pharmaceutical companies have funded many expeditions to tropical regions with the goal of finding plant-derived active substances. About a quarter of prescription drugs are produced from plant materials or are based on chemicals first found in plants, so this strategy makes sense. There are two approaches to plant hunting for pharmaceuticals. One is to collect enough material from an array of plants in a region to test each for chemicals such as alkaloids, a molecular class that includes many drugs. Interesting substances have been discovered this way, but it is rather hit or miss. Baruch Blumberg won the Nobel Prize in physiology and medicine in 1976 for his research in Africa on the viral cause of hepatitis B. To find a treatment for the infection, he led a study that collected plants from three continents and even established an herbarium at the Fox Chase Cancer Center, where he worked. While no useful candidate was discovered, the care with which the research was done is suggested by the creation of an herbarium within a cancer research institute, an unusual venue for one.[17]

The other approach to drug discovery is to partner with indigenous peoples, particularly with healers who use local plants to treat maladies. This has been done since the arrival of the first European explorers, but now the work is much more respectful of local populations. Experts in ethnobotany, often with a background in anthropology as well as botany, live with indigenous peoples and make it a practice to learn the languages of the groups they work with. This is vital to understanding the culture as well as the plants and to studying healing practices and other plant uses.[18] In some locales, for example, palms are pivotal plants employed as food, medicines, building materials, containers, and even cloth, so there is much to investigate.

Abena Osseo-Asare studied efforts to develop drugs from African medicinal plants and discovered that researchers often consulted the herbaria of colonizers to find likely locations for plants that might yield active ingredients. So even though African nations had achieved independence, colonial influence remained. However, she also found it difficult to create a simple narrative of exploitation. Drug development is a complex process, and most African nations do not yet have the infrastructure for research and development independent of multinational corporations. Osseo-Asare's research also revealed that many likely medicines were hardly new to science; their existence had long been known and could not be attributed to a particular indigenous area or group. This meant that compensation would be difficult to negotiate. Robert Voeks has also questioned the "jungle medicine narrative," writing from his perspective as a botanist who spent much time in the tropics studying medicinal plants.[19] He has great respect for indigenous knowledge, but is less positive that useful drugs can arise from these resources.

Unearthing interesting plants with medicinal properties does not often lead to new pharmaceuticals in the United States, where changes to drug laws in the 1960s mandated that a drug's effects must be related to a single substance. This means that most plant materials used medicinally in other cultures would not, as such, be of interest to large pharmaceutical companies. Since plant medicines are often not purified down to a single molecule, their effects are frequently due to combinations of chemicals.

In addition, indigenous healers select plant materials effective for the health problems they encounter, often infections, parasitical diseases, and nutritional deficiencies. These are not the same as health problems in the developed world: cancer, cardiovascular diseases, and neurological problems.[20]

No matter what the aim or outcome, ethnobotanical research requires herbarium vouchers to document the plants discussed in reports and other publications. Having a preserved specimen, a voucher, is a way for future investigators to verify the species tested. Vouchers are also important for work with herbal medicines, essentially formulations of plant material that are not considered drugs but foods by the U.S. Food and Drug Administration. To bring more consistency to this field, it is good practice in their manufacture to voucher each batch of plants used, though "batch" can mean many different things. Ideally, it would be the plants collected in a certain place at one time by a single collector or group working together. This is not always feasible, but it is a worthwhile goal.[21]

Agriculture and Ethnobotany

Food plants as well as medicinal ones are within the scope of ethnobotany. The goal may be to improve food security for indigenous peoples, to better understand their culture, or to explore the history of domestication: how certain plants, such as potatoes, rice, and wheat, became vital food stuffs, and how they were moved from place to place. The genetics of domestication comes in here, that is, the selection of varieties with certain traits such as nutritional value and ease of cultivation. Biogeography also plays a role: how do plants adapt to different habitats? The Russian botanist Nikolai Vavilov was a pioneer in this field and dedicated to improving agriculture in his homeland. He studied plants and collected specimens from many parts of the world to develop his theory that domesticated food plants arose from a few geographic centers, including the Middle East and Central Asia, where various wheat and bean varieties originated. Much of Vavilov's collection of specimens and seeds still exists in Russia, though his work was denigrated in the Soviet Union in the late 1930s.[22]

Vavilov's interest in indigenous crops relates to present-day agricultural research on utilizing native plants as food. In many areas, colonization led in the opposite direction, resulting in great dietary changes. Colonists brought their food crops with them, frequently replacing native crops such as amaranth or millet with rice or corn, species that required much more water and fertilizer than what they replaced and were not sustainable, depleting the soil and succumbing to drought. Exhausted soils often led to turning still more forests into fields. It requires a great deal of study to right these problems, and indigenous knowledge is essential. In the past, this was overlooked, the assumption being that trained scientists knew more than local farmers. The Green Revolution of the 1950s and 1960s began in Mexico with backing from U.S. foundations. They poured money into large-scale agriculture projects that combined high-yielding crops with the machinery, fertilizer, and irrigation needed to support their growth. Early results were often positive but projects were not feasible in the long term. Today's initiatives are smaller, more suited to local needs, and draw on local resources, knowledge, and talents—and local plant species. This does not mean that the work is any less scientific. It is in the ethnobotanical tradition of emphasizing relationships between people and plants, and studying both with teams of botanists, anthropologists, and agriculturalists working together.[23]

Recent ethnobotanical research has moved beyond just searching for medicinal plants or even food plants. There is now a more holistic approach to indigenous plant knowledge, one closely tied not only to finding valuable plants but to saving biological and cultural diversity as well. This makes recording the vernacular names for plants important in field notes and even on herbarium sheets as a way to preserve the knowledge. From early in the history of plant exploration, such information was unevenly respected, though, as discussed earlier, some botanists were careful to record not only common names but local uses for plants. Ethnobotanists are now combing herbarium sheets and field journals for leads on plants growing in their study areas. They are also encouraging *tri-naming* with indigenous, Latin, and common names.[24]

The fact remains, however, that there is a significant gap between the botanical infrastructures of developed and developing nations. Closing what in many cases is a chasm involves more than just sending teams of experts to assist in plant collecting, and sponsoring students to attend graduate schools where large plant collections like those at Kew or NYBG are easily accessible. Infrastructure needs to be developed where the plants are, and this is slowly happening, particularly in countries like Brazil and South Africa, where universities are educating botanists to fill new academic positions and do fieldwork. Those who assist in the field, some called parataxonomists, are receiving instruction in collecting and in plant identification, so those doing the work of collecting are now considered worthy of education and acknowledgment.[25] Visiting botanists can still provide valuable assistance, but they are often as likely to be learning from the permanent staff in local institutions.

Anthropologists and ethnobotanists working together have discovered a close link between extinction of species and extinction of local languages. Research shows that much knowledge about medicinal plants is unique to a particular language; if that language becomes extinct, so does the knowledge described in it. Close observation going on for centuries in indigenous groups resulted in information on plant blooming and fruiting times, plant/animal interactions and, of course, a host of uses for plant material. Ethnobotanists today are helping to document this information with, among other things, herbarium vouchers, to anchor that knowledge to specific data about plants. Indigenous expertise is invaluable but in many cases fading fast since younger generations are often less interested in traditions that might well lead the way forward in environmental conservation.[26]

Throughout the world, herbarium collections are being created by indigenous groups to document the flora that is important to them—a way to preserve plants and the knowledge attached to them. This includes initiatives in developed nations as well. Two indigenous groups in California are working with the University of California, Berkeley, herbarium in making such collections, and the Newe people in Idaho have created an herbarium to document what is growing on their lands. These are manifestations of a desire to maintain a facet of their identity—and a particularly important one, since plants provide so many resources that supported these groups and shaped their cultures. They were influenced by their ecosystems as much if not more than they affected the land, and specimens help to tell this story.[27]

Projects on how language, history, and ecology are woven together can aid those who are not part of such cultures to appreciate that they too are shaped by the plants around them. Just as anthropology has widened its focus and now investigates the activities of scientists and other groups in the developed world, the same move could be helpful in ethnobotany. Since everyone uses plants, it would support the future of botany, and of society, to put more focus on everyone's relationships with plants, if for no other reason than to encourage plant awareness. Appreciation of plant connections might generate some thought about how, for example, buying carrots in a plastic bag provides an impoverished experience of this vegetable. If this is the only way people encounter carrots, they have no idea how beautiful their leaves and flowers are, or how good a fresh carrot tastes. The fact that in the same country where carrots come in plastic bags there are indigenous peoples preserving their biological heritage in herbaria suggests how biocultural issues can vary greatly within a geographical area. Looking at the culture of plant use could make everyone more appreciative of how important plants are to our lives.

Understanding and Conserving Biodiversity

BIODIVERSITY, A TERM THAT HAS BEEN used often in this book, is part of the public discourse on how the changes humans have wrought are threatening the richness of life on Earth. The word became common only in the 1980s, thanks in large part to the writings of Edward O. Wilson, a Harvard University expert on insects, ants in particular.[1] It may not be a coincidence that insects are by far the most diverse group of animals or that Wilson did research in the species-rich tropics. He was alarmed by the destruction of habitats he witnessed there as rainforests were destroyed to make way for monoculture plantations of rubber trees, sugarcane, and other crops. At the same time, many biodiversity "hotspots," such as coastal areas, are being radically changed by development while urbanization is occurring all over the world. These trends are leading to species shifting into new habitats, resulting in some of them becoming invasive and destroying native organisms. Added to these problems is climate change, which results not only in perilous flooding of entire coastal communities and ruinous fires destroying millions of acres but in slower changes to ecosystems that affect species survival.

It is easy to become overwhelmed and pessimistic about conserving species wealth, but the only way forward is to learn more about what is happening and why, and then take action based on that knowledge. An example of the botanical community's efforts is a periodic report published by the Royal Botanic Gardens, Kew, featuring contributions from a global roster of researchers. Called *State of the World's Plants and Fungi,* it is a useful blend of optimism and caution, presenting how biodiversity is catalogued, what is being learned about it, and how it can be conserved and used into the future. It's not surprising that Kew would take a lead in this work with its impressive research staff and the world's largest herbarium. It has a long history of studying diversity in developing nations—though, granted, for a good portion of this history Kew's efforts were on behalf of the world's largest colonial power. As noted in earlier chapters, garden administrators directed far-flung collectors who relied on the expertise and labor of countless indigenous assistants and enslaved persons in finding plants that often proved profitable to Britain. The past and future are intertwined at Kew in complex ways in its living and preserved plant collections.[2]

Cataloguing Biodiversity

According to the Kew report, 1,942 new vascular plant species were described in 2019. Vascular plants have tissues that conduct water and nutrients; they include ferns, gymnosperms such as pines, and flowering plants, but exclude green algae as well as mosses and their relatives. There is no solid estimate of the Earth's total plant diversity, of how many different species exist. The best record of known plants runs to around 450,000, with over 340,000 flowering plants. The situation with fungi is cloudier. That the latest Kew report includes fungi suggests an appreciation of how important these organisms are, though their functions are relatively hidden as agents of decay and nutrient transfer in and on plant roots; there are also many fungi that interact with other organisms. There are almost 150,000 named and described fungal species, with 1,886 added in 2019. As more research is done on fungi, greater diversity is becoming apparent; estimates of fungal species numbers now range from over 2 million to nearly 4 million.

Large collections of fungi and lichen in many herbaria are sure to harbor as yet unidentified species. Today's greater focus on fungi arose from research on how they not only cause many plant diseases but are essential to the life of most plants. Fungi range from microscopic one-celled species such as yeast to familiar mushrooms, which are the reproductive structures for organisms that are made up of fibrous strands called hyphae that usually spread beneath ground through the soil. Some fungi grow on plant roots; others infiltrate root cells. They are instrumental in absorbing nutrients from the soil and making them available to plants, as well as conducting substances among individuals and even species in a habitat.[3]

Examining old plants in new ways, botanists recently looked to the roots of herbarium specimens to identify a species' fungal partners and have successfully extracted DNA from them. Also, the soil on roots can harbor algae, yet another organism in a vascular plant's ecosystem—and a reason to leave a little soil on a specimen's roots, though this is considered by some haphazard specimen preparation. Also being investigated is insect damage to leaves using a grid system to calculate the extent of eaten areas, and it is not uncommon to find dead insects on a specimen and even tiny snails hiding out. So herbaria can be sources of many kinds of biodiversity beyond the plant world and contribute to ecological studies on multispecies interactions, including those involving plant pathogens.[4]

The great biodiversity in tropical areas means some countries, though explored for centuries, are still yielding discoveries. Brazil, Madagascar, India, and South Africa have been mentioned frequently here as collection areas from the sixteenth century on. In a number of projects, these sites have been revisited, with older collections used in planning surveys. The new work may re-collect the same specimens, which can be used in genetic comparisons with the older plants. Not finding some species suggests changes in the habitat due to climate change or other factors, and, not surprisingly, there are new species found as well.[5]

Island ecosystems are particularly rich in endemic species found nowhere else: 83 percent of Madagascar's 11,138 native plant species are limited to this island, making learning about and protecting its flora especially important. A recent study in New Guinea reports that it has the world's richest island flora, with 13,634 species, 68 percent endemic. This is the first comprehensive plant list for the island, and the study could be a model for future work in other areas, though it may still be incomplete. There are almost 4,000 tree species on the list, but this is less than half the number found in an Amazon region inventory. Fewer plots were surveyed in New Guinea, which may be the reason for the lower number, an indication that discovering biodiversity is both difficult and not near its end, while these species-rich areas are under increasing threat from development.[6]

Many factors are involved in estimating biodiversity. It is not just the density of sampling that is important, but where the sampling is done. Studies of the geographic locations of herbarium specimens have uncovered many collection biases because botanists, being human, tend to collect relatively close to home. On expeditions, some areas are easier to access than others. For former colonies, the regions around botanic gardens were often well studied, as were supply routes, nearby seaports, and other urban areas. South Africa, a species-rich region, was explored from the seventeenth century, but there were no collections in some locales until the end of the nineteenth century when they were opened to agriculture. Few collections were made in highly diverse portions of Cameroon until they were surveyed over a decade beginning in 2004; 2,240 plant species were found, about one-tenth of them under threat of extinction.[7]

Though the emphasis here is obviously on herbaria and plants, biodiversity has to be looked at holistically since it extends throughout the evolutionary tree of life. While there are at least nine thousand bird species and about sixty-five hundred mammals, there are approximately 1 million insect species, and there are probably as many more that have yet to be named. The herbarium of the Natural History Museum, London, has three hundred thousand specimens of one-celled algae called diatoms that include at least twenty thousand different species, to cite just one small branch of the tree. Yet it is not just a matter of identifying species; that's only the first small step. A much larger job is learning about them, their life histories, and their interactions. Species do not exist in the world the way they do in collections, as neatly organized individuals. They live with, on, and inside each other, and the life of each species is affected by, and in turn affects, many others.

Learning about Biodiversity

Herbaria are essential reference libraries for learning about biodiversity. A group of biologists analyzed over 13,700 research articles published between 1923 and 2017 that relate to herbarium collections. Among topic frequencies, taxonomic work ranked highest throughout the study period. However, research approaches using herbarium specimens are broadening. Not surprisingly, newer areas include DNA sequencing of specimens and tracking invasive species over time along with other measures of environmental change. The analysis provides the herbarium community with solid evidence of the continuing, and growing, importance of collections. Yet in terms of the uses discussed, it does not go much beyond those in one of the field's classic papers, Smithsonian systematist Vicki Funk's "100 Uses for an Herbarium (Well, at Least 72)." She thought expansively, listing everything from employing a specimen as a reference in identifying a plant to teaching botany and providing inspiration for artists.[8]

While herbaria harbor biodiversity, the term has several layers of meaning. Usually it refers to species diversity, but it can also denote ecosystems with high species numbers, called biodiversity hotspots. In these, a richer variety of interactions among species can mean a more stable ecosystem, or at least one more resilient in the face of disturbances. Then there is the genetic diversity within a species, often a measure of a species' ability to adapt to environmental change. Usually widespread, populous species have greater genetic diversity than those with narrow ranges and small numbers. In order to get some sense of a species' genetic variability, it is important to have a number of herbarium specimens for any one species, both for DNA sequencing and studies on traits.

One or two specimens may be enough to detect hypervariable DNA sequences in vascular plants used to identify species. These sequences are termed a "barcode." Like the type of barcode used to scan purchases, it's a unique identifier, a few short sequences that are distinctive for each species. Barcoding provides a way to distinguish among plants genetically without having to sequence larger portions of their genomes, a boost for those attempting to catalogue biodiversity. The data can be used in working out evolutionary relationships among species and in identifying difficult to distinguish species.[9]

DNA sequencing makes specimens valuable in phylogenetic research to understand evolutionary relationships among species. Using herbarium specimens from historical collections as well as recently collected potato specimens, researchers determined the origin of European potatoes from Andean varieties, later interbred with cultivated forms from Chile. As to the *Phytophthora* pathogen strain that caused the potato famine in Ireland in the 1840s, herbarium specimens revealed that it originated in Central Mexico rather than in the Andes as previously thought. More recently, sensitive DNA sequencing techniques detected citrus bacterial pathogens in plants. There is even military interest in such research. The Center for the Study of Weapons of Mass Destruction in Washington, DC, issued a report that outlined why natural history collections, particularly specimens containing toxic substances, can furnish information needed for protecting against biological warfare.[10]

One reason for renewed interest in natural history collections in general is that they provide evidence of climate change. Since increased atmospheric carbon dioxide (CO_2) is one cause of global warming, botanists have studied stomata, pores on a leaf's surface that control the movement of air into the tissue, and thus of the CO_2 necessary for photosynthesis. New Zealand researchers counted stomata on herbarium specimen leaves, some dating back to Captain Cook's first voyage to New Zealand in 1769–70. There was little difference in stomatal density between eighteenth- and nineteenth-century specimens, but recently harvested leaves had 50 percent fewer pores, perhaps resulting from the increased CO_2 concentrations in the air allowing the plants to absorb the same amount of gas while creating fewer of these structures.[11]

Another way to detect the effects of climate change is with phenology, studying the timing of specific events in a plant's life cycle, such as first leaves, flowering, or fruiting. These visible traits are sometimes obvious on herbarium specimens and have been measured in many studies over the past decade. While there are cases of phenological changes over time, no clear cause and effect relationship exists between increasing temperature and phenology across species. Complicated dynamics are at play, so a variety of species in different habitats need to be investigated. Some species seem more affected than others, and some show little effect; these differences involve multiple factors. Species with flowering controlled by temperature are more likely to be affected than those with flowering triggered by day length.

In semi-arid ecosystems such as a short-grass steppe, temperature influenced flowering in spring-blooming species, but for those flowering in the fall, rainfall was the major factor.[12]

Mismatches in response to climate change can disrupt relationships among species. A Japanese study showed that a spring plant blossomed earlier as a result of earlier snow melts, but its insect pollinators did not appear earlier. This disconnect in timing meant the plants were less likely to set seed, and the insects, when they did emerge, were deprived of a food source that was available in the past. Not all studies find significant differences over time. One compared leaf hair lengths, using fresh leaves and those on older herbarium specimens. The hypothesis was that increasing UV-B radiation due to destruction of the ozone layer would lead to longer hairs providing more protection from radiation. When plants from the same invasive species at various latitudes and thus with exposure to different levels of radiation were studied, no significant difference was found.[13]

There are many other ways specimens can be used to learn about plants and their attributes. Chemists as well as botanists mine herbaria for new data: to search for substances with medicinal effects, plant metabolites produced only during insect infestation, heavy metal pollutants and, of course, DNA for genetic studies. In each case, researchers have carefully tested historical specimens, both to verify that these substances persist and to reveal that variations over time can tell tales of pest invasions and changes in pollutant levels. In the case of metals, scientists are searching for plants that accumulate large amounts of a metal, such as species growing in soils with high nickel levels. These plants might someday be grown to "mine" the metal from soil or to reduce toxic metal levels.[14]

Conserving Biodiversity

A prime reason for learning about biodiversity is to find ways to conserve it. About one-third of the world's plants are threatened with extinction. Such loss would be comparable to the mass extinction 65 million years ago that brought the age of dinosaurs to an end. But it was not just dinosaurs that disappeared; that's not the way the living world works. Species are so interconnected that a single extinction can result in greater loss; the species' parasites and those that preyed on it could also suffer population crashes leading to extinction. That's why conservationists focus on protecting habitats and ecosystems, not individual species such as a beautiful orchid or bird. Species that catch the human eye are linked to many others: microscopic invertebrates and fungi that are less obvious but equally important.

One of the outgrowths of the Convention on Biological Diversity discussed in chapter 13 was the "Global Strategy for Plant Conservation" of 2002. It has five objectives and within them sixteen goals or targets, updated in 2011 with more defined outcomes. While not all the targets have been met, much work has been done on them, and botanic gardens as well as herbaria have been at the forefront of these efforts. In fact, Botanic Gardens Conservation International, a global partnership, has taken a lead. Not surprisingly, the first objective was to understand plant diversity, and one of the targets was an online flora of all known plants. Though not yet complete, it includes substantial information and links to where more can be found. One response to the target was a massive project involving almost a thousand Brazilian botanists and resulting in an online Brazilian flora with information on over fifty thousand species.[15]

Another target was to determine, "as far as possible," the conservation status of all known plants. This goal is more difficult to achieve, but the International Union for Conservation of Nature (IUCN) Red List of Threatened Species, established in 1964, is the most comprehensive inventory of species vulnerability worldwide, though rare plants are less likely to be included.[16] Organisms with economic or cultural value, and those most apparent to humans, are listed more frequently. In the case of plants, evaluation includes research in herbarium collections to determine a plant's range in the past compared to what it is now, and at times to clarify precisely the species being listed.

The second "Global Strategy" objective is conservation and the goals are multispecies in scope, including conserving 15 percent of each ecological region or vegetation type and 75 percent of the most diverse areas, the biodiversity hotspots. Other goals include growing threatened plant species *in situ,* that is, in their natural habitat, and also *ex situ* in botanical gardens or other protected areas, preferably in the plant's home country. This last goal highlights the importance of botanic gardens in biodiversity conservation. They have become safe havens for many species, where plants can be cultivated and propagated. This is an outgrowth of the mission of colonial botanic gardens and their mother gardens in Europe, such as Kew in London and the Jardin des Plantes in Paris. Now there are more gardens in plants' home countries and supervised by the countries' citizens who, thanks to the CBD, can make binding decisions about how these plants will be distributed and used, including for ecosystem restoration projects. However, the resources of gardens in developed countries are still necessary; they nurture plants on-site and send them back to their countries of origin for rebuilding habitats.[17]

Botanical gardens are also preserving specimens to document what is grown and saving small samples of plant material under ultra-cold conditions for future use in DNA sequencing studies. Ideally, each sample is associated with an herbarium specimen voucher for reference. Just as herbaria were important to economic botany in the nineteenth century, they are crucial to biodiversity research in the twenty-first. The number of herbaria in developing nations has increased substantially in this century and the specimens housed there are rising at an even faster pace. The best resource for tracking herbaria and their growth is the annual report of the online *Index Herbariorum,* the definitive guide to the location and holdings of the world's herbaria.[18]

The third strategy is for plant diversity to be used in a sustainable and equitable manner. This depends on support from herbaria and from the international community implementing the CITES treaty to prevent trade in endangered species. Economic deprivation in many developing countries exerts great pressure to profit from natural resources, including plants and animals. This can be through clearing land for agriculture or uprooting rare plants for sale, which does more than just reduce the population—it damages the entire habitat and makes it less likely that the plants can grow back. Herbaria frequently provide evidence in court cases arising from the seizure of endangered plants. Taxonomists may be called on to verify that the species is in fact on the Red List. The same is true of illegal trade in timber from endangered tree species, with xylaria collections' wood samples compared to recovered timber.[19] CITES does have a downside as far as herbaria are concerned in terms of shipping specimens internationally. Forms must be submitted to certify that the material is only for research purposes and not for profit. At times the red tape can make herbarium managers' work frustrating.

Using Biodiversity

The "Global Strategy" targets deal not only with conserving plant species but with using them. The rising human population exacerbates environmental problems and the demand for resources. *Sustainability* is a term that suggests a solution: employing resources in a way that can be stably continued over time and renewed. Many of the themes already covered in this book come into play in this effort, from saving seeds to using plants' genetic diversity and the rich plant knowledge of indigenous peoples. Seeds have always been of interest to botanists as easily transported plant packets: a way to share with the potential to produce future generations in great numbers. Luca Ghini did not just create an early herbarium, he also kept a catalogue of the seeds he collected from plants at the botanical garden he founded at Pisa. He sent the list to other botanists and offered them seeds of any listed species, though seed saving was going on long before that. Farmers kept seed to plant the next year's crops, taking those from the best performing plants, thus selectively breeding for particular traits.

As agriculture scaled up and became more mechanized, a different model developed, with farmers buying seed from companies that grew plants for seeds, often with limited genetic variation. Recently, seeds for many crops are from genetically engineered plants with traits like increased nutrient levels, resistance to pests, or faster growth. Using these seeds decreases genetic variation in crop plants, with resulting susceptibility to new pathogens; with greater genetic diversity, at least a portion of the plants would survive an infestation. However, some farmers and gardeners saved seed from what are called heirloom varieties or landraces, strains developed to grow well in particular areas. These growers were doing a service to the larger community by conserving and propagating biodiversity and are now more appreciated.

Many herbaria have seed collections; they were popular early in the century, and were often sold in custom-made cases, with each seed type in a labeled vial. These samples were not meant for propagation—seeds usually lose their viability rather quickly—but for reference in identifying species collected with seeds. In most cases the seeds are so old that they will not germinate, though they can be a source of DNA and, in rare cases, they have been coaxed to germinate, as have seeds taken from herbarium sheets, meaning that at least a few extinct species may return.[20]

Seed banks, on the other hand, are designed to save seed for future planting, and some are of long standing. They are crucial in preserving the genetic diversity of crop plants, their wild relatives, and also plant biodiversity in general. Many nations have seed banks, especially for agricultural crops and also for horticulturally important species. The massive Svalbard Global Seed Bank built into permafrost within the Arctic Circle focuses on crop species, including landraces and crop wild relatives that may have traits that could be incorporated into new crop varieties. This bank was created as a backup facility for seed collections throughout the world, in case any suffer damage. The largest seed repository is the Millennium Seed Bank managed by Kew that aims to store seed for as many plant species as possible, so it collects broadly.[21]

There are other types of agricultural collections as well. At Germany's Max Planck Institute for Plant Breeding Research, the Ross Potato Herbarium was founded after its namesake collected specimens as well as seed potatoes in South America in 1959, and it grew as additional material was added. In the United States, many crop-related specimens are housed in the institutions that grew out of the nineteenth-century land grant colleges. Their herbaria often have large collections of cultivated plant specimens because of their strong horticulture and agriculture programs. The University of California at Davis is known for these and, being in California, it had a viticultural herbarium of grape vine specimens that has now been incorporated into the general herbarium.[22]

Many of these institutions also have impressive fungal collections since fungal diseases can be so damaging to plants. There is an infamous story about a collection of forty thousand specimens of a plant-infecting fungal group called rusts. This was amassed by Joseph Arthur during his years on the faculty at Purdue University. Arthur considered the collection his own because he had spent his money in developing it, but the university thought differently. At his retirement, Arthur hired a moving van and managed to relocate the entire collection, including cabinets, to his home. The university's president was not amused and began legal action against him. Negotiations ensued, Arthur was reimbursed for his expenses, and the rust specimens—in their cabinets—went back to the Purdue herbarium, where they remain today.[23]

Cultivated plants are the focus of Britain's Royal Horticultural Society herbarium, which includes specimens of plants selected in RHS breeding trials. What makes their sheets distinctive is that photographs are often included, the flowers dissected, and reference is made to the RHS color chart to document the precise colors in the living plant (fig. 14.1). Cornell University's College of Agriculture was for many years headed by Liberty Hyde Bailey (1858–1954), who called his personal herbarium a "hortorium" because of its focus on cultivated plants, *hortus* being the Latin for garden. To this day at what is now the Cornell Hortorium Herbarium, cultivated plants are kept in separate folders, as they are at many institutions with extensive collections of cultivated plants. Even in herbaria with few such specimens, labels note cultivated plants. This is especially important for non-native species, to discriminate between those that have moved into new territory, perhaps as invasives, and those purposefully planted. At times the distinction is difficult to make because some invasive plants were sold as garden plants that then spread out of control. Purple loosestrife, *Lythrum salicaria,* is one example in the United States; it has beautiful flowers and made an attractive addition to flower beds, but now has become a noxious weed—its sale is banned in most states.[24]

Some herbaria, particularly those with ties to indigenous peoples and to the high-diversity areas where many reside, can be particularly focused on species that have agricultural and medicinal uses. These communities

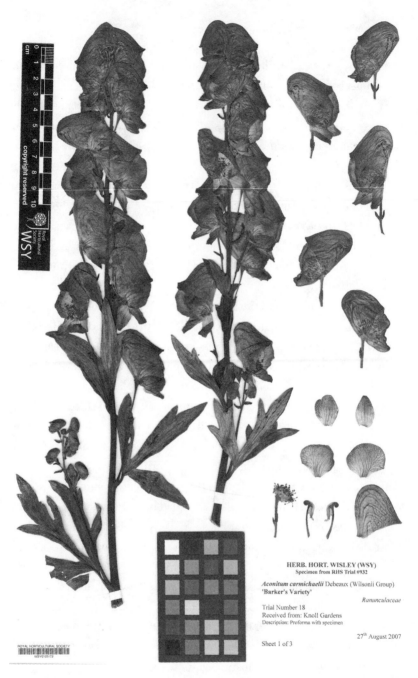

Figure 14.1. A monkshood (*Aconitum carmichaelii* Debeaux [Wilsonii Group]), "Barker's Variety" specimen in the Royal Horticultural Society herbarium; includes individual flowers and flower parts. (The 1851 Royal Commission Herbarium, RHS Garden Wisley)

are also the source of many plant varieties now of interest as landraces grown for generations outside of the agricultural-industrial complex. It makes sense that if biodiversity is important for sustainable agriculture, then working with local communities is vital, as has been done for many years in collecting potato varieties in the Andes for the International Potato Center in Lima, Peru. This center and others around the world that focus on specific crops, such as rice, wheat, and corn, not only store valuable genetic material but do research on plant strains with increased nutrition and other useful characteristics. They also work with local groups in finding ways to make agriculture sustainable. There are efforts to move away from concentrating on a single crop, the result of colonial policies, and to create agricultural practices less damaging to the soil and the surrounding ecosystem. Mixing crops, including within forest environments instead of completely cutting down trees, is becoming more common and definitely in line with the "Global Strategy" and with preserving cultural and biological diversity.[25]

Biodiversity is a multifaceted concept and must be approached from many different perspectives, from local to global. This complexity requires sophisticated handling of information and coordination. The next chapter will deal with some of the powerful new tools of the digital age used to organize information about biodiversity and make data more readily accessible and usable globally.

FIFTEEN

Online Herbaria

NATURAL HISTORY COLLECTIONS ARE essential to learning more about the earth's biodiversity and how it can be conserved and also utilized. One specimen can help answer several research questions about species range, genetics, and morphology. Yet it's impossible to predict how specimens might be employed in the future, as new techniques develop and new questions arise. In the twenty-first century, the realization of collections' growing value resulted in massive efforts to image specimens and enter their label information into databases, making them available to a broader community of users via the internet.[1] Pressed plants are the easiest to deal with because they are primarily of standard size and two-dimensional, though there were still many technological hurdles. The complex process of digitization pushes hardware and software to their limits, yet it's among the most exciting aspects of biodiversity research, spurring new uses for collections as not only taxonomists but ecologists, chemists, conservationists, and historians learn to mine the data.

The Road to Digital

In 1960, if a researcher wanted to look at specimens of a particular species, she or he would have to visit several herbaria or ask to borrow sheets. A curator might have to hunt for the requested specimens because herbaria, while resembling libraries in some ways, did not necessarily catalogue and shelve every specimen that arrived. Like libraries, some herbaria had card files for their specimens, others a ledger system, but some did not keep up-to-date records. New acquisitions, which were usually unmounted, were often simply stored for future attention—sometimes for years, even for a century or more. Mounted specimens might also be difficult to locate, since the species name could have changed yet the specimen was still filed under the old name. In addition, an inquiring botanist would need to search the literature for information about the species, including the publication where it was first described. This might be in a two-hundred-year-old journal, perhaps in a foreign language. Locating publications would mean using various finding aids, all in paper format and often themselves not readily available.

This situation began to change, slowly at first, in the 1960s when computer databases became workable. In a prescient 1969 article on herbaria of the past, present, and future, Stanwyn Shetler, then curator at the United States National Herbarium of the Smithsonian, used statistics from the latest edition of *Index Herbariorum* to survey the state of herbaria worldwide. At the time, 1,850 were listed, containing almost 150 million specimens. Because of herbaria's massive resources, including centuries-old material, Shetler argued for computerized information retrieval systems. He foresaw inputting label data, which could then be searched for a particular species, collector, or location.[2]

Four years later, Shetler reported on a pilot project to create an electronic register of type specimens held in major U.S. herbaria. Using a large genus as a test case, the plan focused on accuracy and laborious verification. The project was soon subsumed into more flexible programs, but its flaws were valuable in showing the way forward: what would and would not work. Shetler suffered the curse of the pioneer: those in the technological vanguard had to redo his work as systems become more sophisticated. In the same year as Shetler's report, Howard Irwin of the New York Botanical Garden headed a committee that published a national plan for America's natural history collections with a focus on retrieval systems.[3] However, with cell and molecular biology producing dazzling results and leading research away from systematics, resources were not available for a project on the scale Irwin outlined.

Digital Resources

By the end of the twentieth century, individual herbaria were digitizing label information and creating searchable databases made available on the Web. Some institutions were even imaging specimens, either by scanning or photographing them. Often botanists built on the work librarians were doing in digitizing card catalogue data and scanning books: flat herbarium sheets could be treated much like pages of text. As the work progressed, problems of scale cropped up and broader possibilities for botanical databases appeared on the horizon. Addressing these would require collaborative efforts by the natural history collection community as a whole.

One project focused on digitizing botanical type specimens. As mentioned earlier, most types, even of tropical plants native to developing nations, are held in European and North American herbaria. To make these specimens available to researchers worldwide, the Andrew W. Mellon Foundation began the African Plants Initiative in the early twenty-first century to digitize species endemic to a continent with relatively few active herbaria.[4] However, the specimens would be of little use without relevant literature. Mellon partnered with the nonprofit digital library JSTOR to scan botanical literature and create a website where searching for a particular species would call up not only type specimens but links to related literature and, in some cases, illustrations that might have served as types. Ultimately, JSTOR fashioned a massive portal, that is, a website providing information from many sources as well as tools for using the information and links to still other resources.

Mellon furnished high-resolution scanners and training to herbaria with type collections; eventually the project expanded beyond African species to those from Latin America and then throughout the world. The endeavor is now complete, and the Global Plants portal delivers access to images of over 2 million type specimens, thousands of illustrations, related literature, and even the digitized correspondence of such noted botanists as Joseph Dalton Hooker.[5] In order to maintain the site and keep it functioning, JSTOR charges for access by individuals and institutions. Though funding makes the site available to researchers in developing nations, it does not solve continuing problems with technology there.

As ever more scientific inquiry relies on virtual data storage and analysis, open access becomes a larger issue, yet it has been lurking since the 1960s, when DNA sequences began to be databased. There were attempts to charge for this information, but ultimately the natural history museum model was used. Sequences were treated much as herbarium or zoological specimens are: made available to any researcher who wanted to use them to build on knowledge in the discipline. The same model is now widely employed for sharing specimen data. The Global Biodiversity Information Facility (GBIF) is composed of national nodes in dozens of countries, each contributing to building and sustaining an infrastructure that stores well over a billion organism records, both specimens and observations.[6] There are many other such projects at the national and international levels.

For printed literature, Biodiversity Heritage Library (BHL) began as a consortium of U.S. and British libraries, with the Smithsonian Libraries assuming administrative responsibility. The number of institutions supplying digitized publications and manuscripts has steadily grown and become more international. The BHL collection consists primarily of material published before stricter copyright laws came into effect in 1923. Historical sources are important to botanists, because priority is given to the earliest published description of a species. In addition, many scientific organizations contribute more recent publications. Also available through BHL are manuscript collections, such as the correspondence of Asa Gray and John Torrey, as well as field notes, helpful in understanding naturalists' collection patterns. To attract a larger audience, BHL created a Flickr site with over 250,000 illustrations from the books and journals it had scanned. This has proved a powerful tool in making scientific illustrations more accessible, especially to nonscientific audiences.[7]

In imaging printed matter and specimens, the issue of the tangible versus the virtual arises. Though many specimens were scanned with equipment also used to scan books, these items are different from each other. It might at times be possible for a library to dispose of a printed copy of a book, but books do not have DNA, interesting chemicals, and other characteristics that don't come through on a screen. Digital images are not replacements for specimens; the book analogy doesn't work here. Nor is a specimen two-dimensional like a printed page; it has depth and texture that becomes less obvious in an image. Specimen images are usually of high resolution so they can be magnified for a closer look at details. Still, this is not comparable to putting a specimen under a dissecting microscope as botanists do routinely. There is also loss of scale and context—the herbarium, the cabinet, the folder containing it and other specimens. Nonetheless, there are also advantages. Up to now, it has been difficult to put specimens from different collections next to each other without the formality of borrowing specimens. New image presentation technologies, first developed for the art museum community, are being adapted for natural history collections to make this possible.[8]

Large-Scale Digitization

A daring approach to preserving collections, digitizing them, and planning for their future use was taken in the Netherlands, where all the natural history collections, relevant literature, and archives were centralized in one place, the Naturalis Biodiversity Center in Leiden with a newly built facility. This massive undertaking was not without problems, beginning with protests from institutions that were losing materials and thus easy availability. Many collections dated back hundreds of years and were housed in historically significant facilities, though ones that admittedly needed updating or more. With the completion of the transfer in 2019, it was apparent that Naturalis would be an important biodiversity research hub not only for the Netherlands but for the European Union and the world, one reason being the large-scale digitization done in concert with the move.

In the United States, the aim became to unite specimens digitally, not physically. As the Mellon project progressed, the National Science Foundation (NSF) funded an initial study to determine what was needed to digitize private natural history collections in the United States, and how these needs could best be addressed; governmental collections like the Smithsonian Institution's were provided with other funding sources. The result was a massive program, Advancing Digitization of Biodiversity Collections (ADBC), awarded multi-million-dollar funding annually for ten

years. ADBC created Integrated Digitized Biocollections (iDigBio) as a national resource for this endeavor, overseeing projects to image and digitize
the data for particular portions of botanical and zoological collections
spread over many institutions. Each project addressed research questions,
such as how plants interact with herbivores that feed on them and how to
measure phenological change in a biodiversity hotspot.[9] iDigBio directed
the development of a Web portal for all this data as well as training programs for collections staff and new tools for organizing the data and making
them useable.

Many iDigBio projects are now complete and the data for millions of
specimens are available through its website and through regional portals
such as SERNEC, for herbaria large and small in the Southeast, and SEI-
Net for the Southwest.[10] Throughout this process, iDigBio has been assem-
bling examples of best practices to speed digitization efforts, such as hiring
experts in industrial workflows to design the most efficient sequence of
steps for digitization. Specimens have to be located, barcoded for retrieval,
and imaged, and a digital record having at least the species name and perhaps more label information must be created.

Though scanning produced extremely high-resolution images, it was
time consuming, often taking several minutes per scan. High-resolution
photography is faster and more than satisfactory. In the early 2000s, the
herbarium at the National Museum of Natural History in Paris, which was
being renovated, took the unprecedented step of packing up most of its
6 million vascular plant specimens and shipping them to a warehouse with
a conveyor belt assembly line to image specimens at a series of work stations. Since then other large collections, including that of the Naturalis Biodiversity Center, have adopted the approach. At the same time iDigBio
made a special effort to include smaller herbaria in digitization projects to
capture the greatest breadth of specimens. These collections often hold
species, frequently collected locally, that are not necessarily in larger institutions with herbaria that may appear encyclopedic.[11]

While processing physical specimens was proceeding, those in the
computer field were improving databases for more effective access to label
information. In order for data to be useful they must be accompanied by

metadata: data about the data. If users want to search an herbarium database for a particular plant—say, the Venus flytrap—they may be disappointed when they type this in the search box, since the software may recognize only scientific names, in this case, *Dionaea muscipula*. If users type in this scientific name, links to specimens of this plant will appear because when that name was entered into the database, it was linked to metadata for "species" name.

Database software entry forms have fields, or blanks, for each type of information, so entry is controlled, and behind the scenes metadata are added (fig. 15.1). Most botanical databases have many genera and species names already stored, so when someone begins typing a scientific name, a list of possibilities displays and one simply chooses the correct one. This avoids the problem of spelling errors, which plagued pre-computerized botany; names were often misspelled in print, thus perpetuating errors. There is even an online service to reconcile nomenclatural issues arising from such mistakes.[12] In addition, if a user tries to enter a plant name in the collector field, the program will not permit it.

Metadata are hardly unique to science; they are needed to make any form of information searchable online. For standardization, there are agreed-upon definitions for certain terms or fields in a database, like name and date, that make up what is called the Dublin Core; this vocabulary enables data publishers to share their information using this common terminology. Because biodiversity research requires additional terms, such as to designate the institution owning a specimen, there is the Darwin Core. This peek behind the scenes of software engineering will end here; it was presented as a reminder that there are difficult issues in what is called the "back end" of software design of which the user is unaware. If a website or computer program doesn't function quite as desired, there is probably a good reason for that. It may be fixable but might take so much time and reprogramming that it's just not feasible to do so.[13]

These issues are significant because iDigBio is charged with broadening the user base for natural history collections. Software that is easy for botanists to navigate in searching for particular species or specimens can prove daunting to a layperson. A search page with no instructions is not

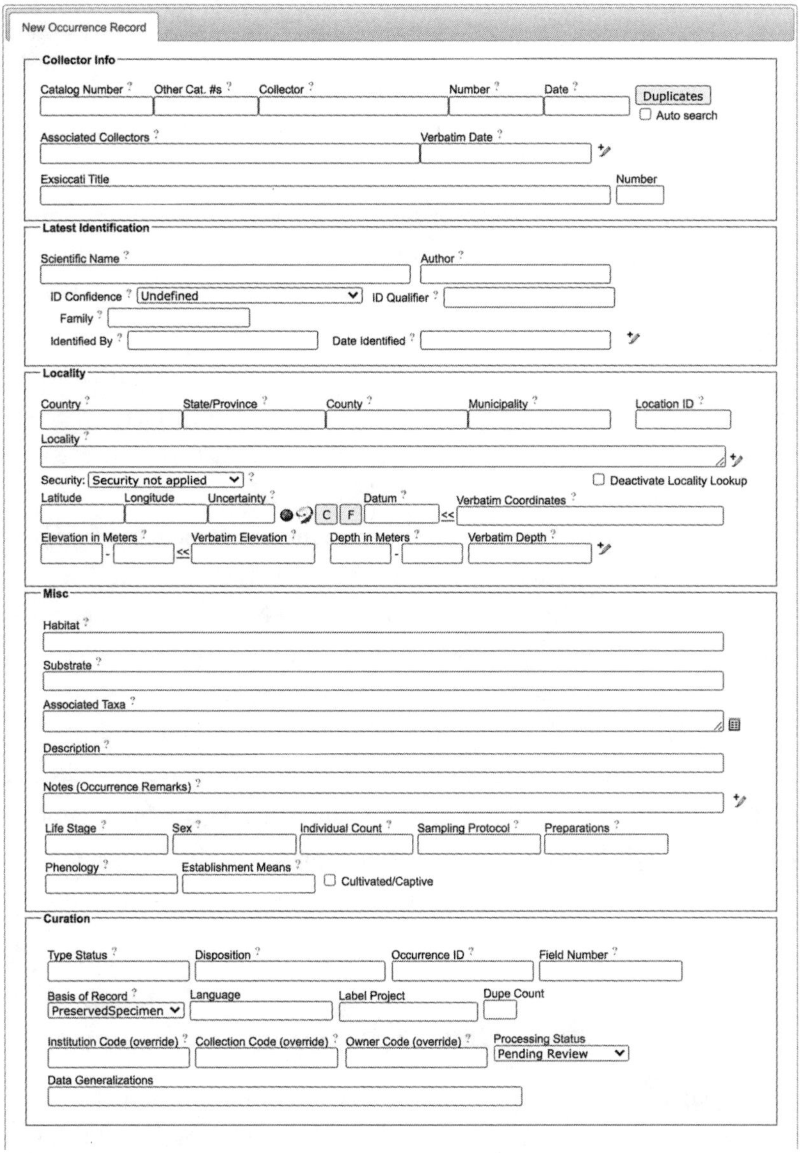

Figure 15.1. Symbiota basic data entry form showing the different fields (boxes), many of which accept only specific information, such as the scientific name of the specimen and the collector's name. (Symbiota Software Platform)

user-friendly for a novice. A worldwide consortium of botanical institutions is developing World Flora Online. This website is attractive and provides a great deal of information, but the search feature is designed primarily for scientific names, very useful for botanists. Digital format is the future for floras in general because they can be updated more easily and stay current, while bound volumes quickly need revision. However, to make the information more accessible to nonspecialists requires editing and reformatting information. There are sites that do this, especially ones hosted by botanic gardens and geared to the needs of gardeners and students.[14]

Collaborations like the World Flora Online are growing because climate change and biodiversity loss are global issues with many interconnecting strands. Yet dealing with data on an international scale entails complex issues, many yet to be fully addressed even locally. iDigBio handles information only from U.S. institutions; there is Canadensys for Canadian natural history collections, the Environmental Information Reference Center in Brazil, the Atlas of Living Australia, the National Specimen Information Infrastructure in China, and Distributed Systems of Scientific Collections (DiSSCo) for Europe.[15] The goal is to have these separate entities interconnect. To a certain extent this is achieved by GBIF, which is an aggregator—that is, it accepts data from many databases *if* they are in a suitable format. In reality this means that data from iDigBio, for example, is uploaded to GBIF, but not all information fields may be included. Compromises like this are necessary at the moment, but there are projects planned for more data integration in the future.

Digitization 2.0

While iDigBio has created a massive resource, each new achievement brought visions of other possibilities and each new tool revealed previously unseen technical issues that needed to be addressed, such as how diverse computer systems could be integrated. This led to planning for the next steps, with similar work going on with Europe's DiSSCo. There are still many specimens that need to be digitized, but the emphasis now is on what some term "Digitization 2.0": how data can be used, how to develop the user base, and how to create tools that will answer ever more sophisticated questions about the living world.[16]

One proposed concept, growing out of iDigBio, is the Extended Specimen Network (ESN), which has three layers of connections out from the physical specimen itself, which would be at the network's core (fig. 15.2). The first level includes the digital specimen record and any accompanying specimen images, including, in some cases, a 3-D image file. The next layer has links to field notes and images, gene sequences, morphological measurements, and isotope data. The third contains phylogenies or evolutionary relationships, species descriptions, ecological interactions, distribution maps, and conservation data. While ESN requires a massive infrastructure, the vision of expanding an herbarium sheet is hardly new. In 1952, Edgar Anderson wrote that "making a good herbarium record . . . is something like trying to stable a camel in a dog kennel" to describe his frustration in trying to capture all the features of a plant on one sheet. To document his research on corn, he pasted photographs of tassels along with notes on his findings to herbarium sheets, to be filed in the herbarium along with corn specimens, or, because these plants could be so bulky, with photographs of the entire plant. He first considered this a minor change in record keeping, but as he realized its usefulness and considered its implications for keeping data and specimens together, he dignified it with a name, "the inclusive herbarium."[17] Anderson's work can be seen as an early analogue version of integrating botanical record keeping.

ESN envisions a user entering the name of a species into a portal and from that one website accessing all information about the species. If this concept is realized, it might counteract the increasing specialization in biology that has so changed the discipline over the past two hundred years: geneticists can easily call up information on the source species for a gene sequence or ecologists can find phylogenies for species in a food web, while both access the species' specimens as well. These visions are exciting, but entail questions about necessary metadata, who owns the information, and how different technologies communicate with one another. Dealing with these issues takes time, deep analysis of computer architecture and, of course, money to finance the personnel, equipment, and facilities to make

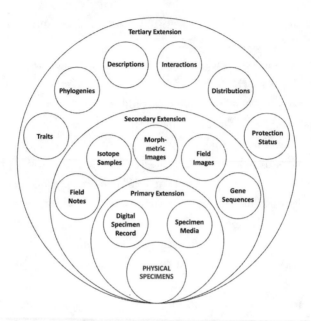

Figure 15.2. Diagram of the extended specimen network concept illustrating the layers of linkage that could be created around an individual specimen. (Adapted from Thiers et al., "Extending U.S. Biodiversity Collections to Promote Research and Education")

integration possible and ensure its future sustainability. Funding is an issue for all these programs, but the European efforts are sustained by relatively stable government financing, while the grant-based system in the United States has not focused on the long-term scale required for this work.

Europe's DiSSCo has proposed something similar to the ESN: the digital specimen. The differences between the two are less about the aims and more about the behind-the-scenes infrastructure. Since ultimately, the goal is to have all specimens worldwide be accessible, it makes sense that iDigBio and DiSSCo are working together, along with the globally oriented GBIF, in the Alliance for Biodiversity Knowledge.[18] The three are moving to develop standards for all collections so they can be integrated. Issues to be worked out include future updating of label information, especially with nomenclatural changes, and how to have the change recorded throughout the system. Also a major concern for institutions is documenting collection use. This is how they justify expansion requests when seeking funding. In the past, a guest book logged in visitors who referenced a collection, and a spreadsheet kept track of loans. In the digital age, it is more difficult to tally use when specimens have been loaded onto large aggregators such as iDig-Bio and GBIF. Such services are now using software that provides contributors with statistics on specimen views and on how often they are cited in publications, an important impact measure.

Though data are input through many different software systems, all need standardized metadata and unifying instructions. FAIR is the acronym for achieving this: making collection data findable, accessible, interoperable, and reusable. Findable means that each piece of information has a unique identifier so there is no ambiguity. This helps to make information accessible from a variety of systems that need to interact with one another: be interoperable.[19] Right now many databases do not "talk" to one another; one cannot draw information from the other, so they operate in separate "silos." Finally, data need to be in a standard format that is reusable when a website is redesigned. There have been wonderful websites presenting the work of noted botanists, with their specimens, manuscripts, and publications; these have disappeared because the software was custom made and not compatible with updated systems.

The ins and outs of these problems are both fascinating and byzantine in their complexity: the goal is to make everything from herbarium specimens to genetic sequences and ecological data available through one entry point. Geneticists, botanists, and ecologists frequently have trouble talking to one another because of their distinct vocabularies, to say nothing of different ways of organizing and labeling data. For example, ecologists interested in plant traits give little attention to herbarium specimens. There are also computer language problems with software interacting, and often the risk of some data being lost in translation. Despite these issues, amazing progress is being made in the digital realm and many specimens are now "born digital," their label data input electronically right in the field at the time of collection. In some cases, additional information is being added to specimen labels to make them useful to a broader audience of biologists.[20]

Using the Data

Creating large-scale biodiversity informatics networks bodes well for the future, and the current infrastructure has already yielded important results. Traditional taxonomic work in describing new species continues, greatly aided by the availability of online specimens, particularly type specimens. This often triggers reexamination of the genus in which the new plant is placed, requiring further study of specimens. The work is now closely tied to cataloguing the Earth's biodiversity, and once published, usually now in online journals, leads to the information making its way into online floras. Another result of digitization is that in many cases, specimens were given updated determination slips when species had been renamed, and the change noted in the database entry.

Data input also frequently involved georeferencing older specimens. Most recent labels provide latitude and longitude for the collection site, along with other locality information, including a place name or distance from it, a brief description of the habitat, and perhaps names of other species found there. Unfortunately, old labels, particularly those from the nineteenth century or earlier, usually have scant locale data, yet these sheets are valuable for tracking changes over time in habitat or the timing of life cycle events, which are often sensitive to climatic conditions. Approximate geographic coordinates for longitude and latitude can be calculated, with a stated margin of error based on how broad an area is involved; for example, it would be greater for the area of a county than for that of a small town. Georeferencing makes specimens more valuable in ecological studies, but it is labor intensive, so this information still has to be added for many specimens.

Ecologists use georeferenced digital records in analyzing changes in a species' range over time and in species distribution modeling: determining areas that might provide suitable habitat for a species based on what is known about its characteristics and ecology. This research is helpful in identifying possible collection areas and also in approximating where plants might be able to grow as the climate changes. However, researchers estimate that twenty to fifty specimens per species are necessary in this work. A survey of specimens available on GBIF found that 90 percent of the species were represented by fewer than twenty collections and so were "essentially invisible to modern modelling."[21] This is definitely an argument for more collecting and more digitization. On a positive note, a study like this would have been impossible before the age of online collections.

Though phenology came up in the last chapter, it also deserves note here because having specimens available online makes studies of flowering and fruiting times much easier. If the presence of flowers is transcribed in a "notes" field, it is more difficult to retrieve this information electronically because this field doesn't have a restricted vocabulary as does a searchable field, which may have only four choices: for example, flower, fruit, both, neither. With the increased interest in phenology and with more sophisticated ways to capture information, this problem is being addressed.[22]

At the moment, data for a majority of the world's specimens have not been digitized, and even fewer have been imaged. It is impossible to come up with an accurate estimate because the field is changing so rapidly. Moving data online is very uneven; some collections are 100 percent digitized and little or nothing has been done with others. Researchers have to be aware of the limitations and biases inherent in collections, particularly in online collections. Besides the collection biases mentioned in the last chapter, there is variability on what precisely digitization means for a particular specimen or collection. In some cases, all that is available online is a "skeletal record" that has the species name and perhaps the country or state where collected. Information on habitat, geographic coordinates, and phenology has yet to be added. Obviously, a record like this would not be of use for many studies.

Although such problems exist, there is exciting new work on herbarium specimens employing artificial intelligence, AI. Computer scientists have used neural network software to simulate deep learning and found that machines could be trained to identify species on herbarium sheets, measure parameters such as leaf area, and even recognize flowers and fruits. At the moment this requires a great deal of computer power and is expensive, but it is evidence that the work will be feasible in the future.[23] AI is also improving optical character recognition (OCR) used in automating label data transcription. OCR works relatively well on typed labels, but less so on handwritten ones; AI augmentation is making the latter more accurate. Also, programs for natural language processing are used in label transcription by identifying particular kinds of information, for example, recognizing when a word or phrase refers to habitat and entering it into the correct field in a database. The potential for AI use in biodiversity collections is significant because with millions of specimens, it would be difficult for humans to effectively extract data in a timely way.

All this technology does not negate the continued need for fieldwork. Biologists argue that in the push to create ever more sophisticated ways to store and use biodiversity data, it would be injudicious not to maintain collecting over time.[24] Resources need to be put into collecting and maintaining specimens physically and digitally. Even though there are over four hundred million recorded herbarium specimens worldwide, these are not enough. Not only are there still new species to be discovered but many areas are undercollected, either because they are inaccessible or are in places where there has not been the infrastructure or economic backing for collecting. Also, areas must be re-collected to discover what changes have been wrought by climate change or human intervention. Herbaria need to keep expanding their physical and digital offerings as well as attracting new users to their world, the subject of the next chapter.

A Broader Vision

Herbaria and Culture

MORE THAN EVER, HERBARIUM CURATORS are realizing the vast potential of their collections and are encouraging involvement by researchers and new users from gardeners to artists. Also, the aesthetic aspects of herbaria, which those in the know have always appreciated, are now being explored and enjoyed by a broader audience. This last trend is a reminder of Michael McCarthy's argument that nurturing aesthetic appreciation of the living world can aid its preservation. In 1973 the botanist Richard Cowan described the herbarium as a data bank, and more recently the artist Victoria Crowe saw it as a botanical memory bank from which she drew inspiration.[1] They are both right, as will become clear in this chapter on the beautiful future for herbaria.

The burgeoning use of herbarium collections, in part the result of specimen digitization, has triggered a host of changes in the botanical community. There are greater efforts to build connections with other scientific disciplines as well as with the humanities, educational institutions, and the general public. Barbara Thiers, former herbarium director at New York Botanical Garden, has written a wonderful introduction to herbaria, and Kelly LaFarge published an herbarium guide for children. There is a new liveliness and creativity among herbarium curators, who have banded together, forming their own society to amplify their influence.[2] With iDigBio and DiSSCo taking the lead, curators are creating strategic plans to further their goals and increase impact.

One factor driving outreach efforts is the need to develop greater botanical expertise for the future. Programs offering degrees specifically in botany are now few and far between, having been replaced by ecological and environmental programs, where a focus on plants may or may not be offered. This means fewer people are trained in systematic work: knowing a particular plant group well and how to find information about the rest. To help reverse the trend, the National Science Foundation funded a project to create an educational framework for the natural history community and future biodiversity research. It developed outcomes for students in fields using natural history collections and described a skill set that includes basic knowledge in zoology and botany as well as familiarity with digital analysis tools for biodiversity informatics.[3]

The focus in biology education still tends toward genetics, human physiology, cells, and molecules, but interaction with the living world and preserving it are becoming powerful themes that, as this project suggests, are gaining ground. In Britain, a concentration in natural history is now offered for high school students, and France's National Museum of Natural History has several outreach programs aimed at students. These former colonial powers are also collaborating with developing nations in supporting biodiversity education and research. Natural history is coming to be associated with the currently popular subject of biodiversity and gaining favor by positive association, particularly in the education community. Perhaps the trend will usher in a "heyday" for natural history in the twenty-first century as there was in the nineteenth.[4]

Curators at college and university herbaria are key to these efforts and are working to lure students into plant biology, including field courses linked to collecting, identifying, and mounting specimens. These classes are hardly new, but with the introduction of social media, educators have novel ways to stimulate interest. The herbarium curator at Virginia Tech has established a natural history society that organizes field trips, and Bucknell University offers a course in which students create art on herbarium sheets, exploring connections between art and botany. The herbarium at the University of Tennessee, Knoxville, sponsors "Specimens and Scones" events using the enticement of food, a sure winner with students. Even better than food is the lure of money. Many herbaria participate in student work-study programs to digitize specimens as well as internships for more extensive work on curating collections. These provide cheap labor, often by biology majors who may become so enthralled that they pursue further studies in biodiversity.

Increasing awareness of collections is also happening earlier in the education pipeline. Museums and botanic gardens are taking the lead in informal education programs. Several institutions have held Harry Potter events introducing children—of all ages—to the plants that play a role in these novels. In a different twist, the Royal Botanical Garden, Edinburgh, provided materials to make "Frankenstein" specimens, imaginary creatures fashioned from pressed plant parts pasted to white paper, embellished with color markers, and labeled with the specimen's name (fig. 16.1). As an up-to-date final touch, each specimen was digitized, so the maker would also have a digital record.[5] Actually creating specimens provides a real, physical connection with nature and makes it difficult to avoid careful observation; drawing a specimen is another great lesson in observation.

Also increasing are more formal school programs in which curators visit schools or small groups of students tour herbaria. Seeing rows and rows of herbarium cabinets filled with specimens might leave a powerful impression that could later lead students to find out if their college has a collection. Even if early introduction to herbaria and specimen-making does not lead to a biology career, awareness that such collections exist and have value could excite engagement with biodiversity. During the COVID shutdown, increased interest in nature led historian of botany Elaine Ayers to start Quarantine Herbarium, a website where she invited people to up-load images of plants they gathered on solitary walks and then pressed; some included drawings. It was a wonderful way to build attention to na-ture and community while making the word *herbarium*—and its meaning—more familiar.[6]

Botanists are also luring adults to herbaria. Tours through collections are now more common, as curators realize that people will develop an interest in herbaria only if they know such collections exist. Most curators have favorite specimens they like to show visitors. It is amazing how many collections, even in the United States, have at least one of Charles Darwin's specimens, often a moss, since they were easy to subdivide into several pieces for distribution. Or

ROYAL BOTANIC GARDEN EDINBURGH (E)

Name: George Hunter
Collector: Zachary   Coll No: 261
Description: purple flower

Locality: UNITED KINGDOM: Scotland: (VC 83): Midlothian: Royal
Botanic Garden Edinburgh
Lat: 55° 57' 50.83" N Long: 3° 12' 42.68" W. Date: 13 April 2019
http://bit.ly/rbgedcsfest

EdSciFest190261

Figure 16.1. "Frankenstein" specimen created by Zach at the Royal Botanic Garden, Edinburgh, on April 13, 2019. (Archives of the Royal Botanic Garden, Edinburgh)

there may be a particularly beautiful flower that has kept its color for a century, or an early record of an invasive species. Curators are also mounting exhibits to show off specimens, at least for a short time, since too much light can damage fragile plant material. The herbarium at the Australian National Botanic Gardens in Canberra has a set of binders with specimens in plastic sleeves. It is available to visitors, interested gardeners, students, and others who just like plants, so they can expand their view of what a plant can look like, alive and dead. These efforts give visitors to gardens and museums a chance to see some of what is stored beyond the exhibits.

Systematists are doing their part by giving new species names that attract attention to their work: a fern genus named for Lady Gaga, who wore a costume that resembled an early stage in fern development, and a tomato in honor of Mark Watney, the title character in *The Martian,* a film featuring an astronaut growing potatoes (same genus) in space. There have even been online auctions in which taxonomists offer to name species after those who pledge financial support to their programs. Through outreach efforts, natural history collections are becoming more visible and valued—Oberlin College, which had sold off its collections years ago, has now begun a new herbarium, as has the Meijer Gardens and Sculpture Park in Michigan.[7]

With grant money from NSF and other sources, many herbaria have acquired added space or at least new cabinets, often as part of moveable systems that increased capacity. As happens in any move, there were surprise finds: treasures that had been forgotten or were unknown to present curators. When the University of Connecticut herbarium relocated, two specimens collected by Henry Thoreau were unearthed and celebrated in the local news as the wonderful discovery it was, and New York Botanical Garden suspects it will add to its known George Washington Carver specimens as it continues to digitize its fungal collection—many have already been found.

The herbarium at Manchester University is housed in a maze of rooms in its museum's attic; it was so packed with books, cabinets, and boxes that it was hard to tell what it held. Now, after several years of renovation, specimens are rehoused in new furnishings and digitized so they are available to the world. Also literally dusted off were boxes of gorgeous botanical illustrations

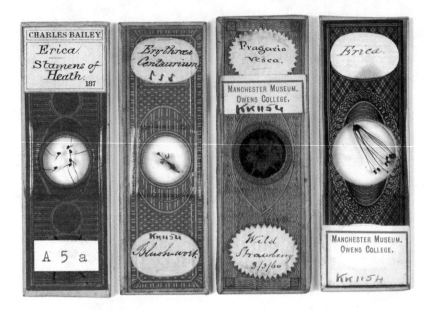

Figure 16.2. Microscope slides of plant material with ornate labels. Originally decorative paper was used to hold the glass coverslip over the specimen in place, but later it was sometimes added simply for aesthetic reasons. (Manchester Museum Herbarium, the University of Manchester)

and plant morphology slides, many with ornate nineteenth-century labels (fig. 16.2). This case is hardly unique. Donna Young at the World Museum, Liverpool's herbarium, has become an expert on nineteenth-century plant models that were ignored for decades and are now valued as artifacts of science and art, and as documents illuminating the history of botanical education.[8]

Public Participation

Herbaria are also involved in citizen or community science: participation by nonscientists in scientific endeavors. These range from advocacy for projects like solar energy development to taking part in environmental legal cases to participating in laboratory research or field work.[9] As far as herbaria are concerned, the most common activities are imaging specimens and entering label data, both labor-intensive and massive tasks, considering the almost 400 million specimens stacked in cabinets worldwide. Obviously, imaging has to be done locally, most commonly by placing each specimen in a light box and photographing it. Once the image is processed and linked to a "skeleton" digital record created with the specimen's scientific name and a unique barcode number, transcribing the label from the digital image can be done online by users anywhere. While optical character recognition can convert the information into electronic data, it works well only with typed labels and even then there are problems. Taking accuracy into account, a human can often input data almost as rapidly as OCR.

Public participation activities are becoming vital to many digitization projects. Yes, people can be hired to do the job, but there are severe spending limits. Interactivity on the Web has improved significantly over the past few years, so that cloud-based software is much easier to use. That's where portals such as Zooniverse come in. It describes itself as "People Powered Research" and hosts projects in many scientific fields. One is Notes from Nature, in which volunteers digitize information on specimen labels. There are online tutorials to get participants up to speed, as well as built-in processes to check accuracy. Curators have expressed surprise at how rapidly projects are completed and need to be restocked with more assignments.[10]

In other online tasks, participants transcribe field notes, correspondence or, in the case of the Biodiversity Heritage Library, add tags to images on its Flickr site to improve searching. Some observers believe that the greatest benefit of these endeavors may not be data entry itself but the transformation brought by people's interaction with digital collections and collaboration to produce new knowledge. With greater engagement in primary sources can come a democratization of research, something that was the case in nineteenth-century natural history, when women on the Great Plains were sending specimens and information to botanists like Asa Gray, and settlers in Australia and New Zealand were corresponding with Joseph Hooker.[11]

Additional projects for those interested in natural history include recording species sightings and other environmental observations with an app such as iNaturalist. This can strengthen interest in nature, and the information is then uploaded to aggregators like GBIF, with over a billion

records available for ecological research. Because specimen collection has
taken a downturn, field observations help to compensate and can be linked
to specimen data, though they cannot replace physical evidence completely.
With iNaturalist, herbaria are now linking observations, including field
photos, to herbarium specimens collected at the same time, a move in the
direction of the extended specimen. There are also apps to help identify
species. Years in development, the Leafsnap app was often able to provide
a name and information about the species from an uploaded leaf photo.
iNaturalist now has a more sophisticated identification tool.[12]

Community science is an old idea—amateurs have been observing
the natural world since the dawn of the human species—but the term has
become a buzzword now for several reasons. Improved technology is obviously one, but there is also the present upsurge in retirees, many of whom
are still young and vigorous enough to seek intellectually interesting activities. Participants can work in herbaria mounting specimens, rearranging collections, or inputting data, and it is great to have help with deciphering
difficult handwriting. At the other end of the spectrum, and more important
for the future, are opportunities for young people to be lured into the natural
history world through volunteering, work-study programs, and internships.

The Humanities and Art

Another significant outreach from natural history collections is to those in
the humanities and arts. Herbaria can be significant historical archives, and
it takes knowledge of history to appreciate the full significance of the specimens collected by Joseph Banks in eighteenth-century Australia or by the
Wilkes Expedition in nineteenth-century California. Having these collections online means they are available not only to botanists but to historians
who can mine label data. In many cases related manuscripts are also being
digitized. These massive undertakings are opening up many new research
questions. For pre-Linnaean herbaria, it means the end to a long period of
neglect because they were considered of little value in taxonomic work.
Now they are seen as troves of information on the horticulture, agriculture,
and even climate of the past.

In a project connecting several British institutions that hold material collected by Hans Sloane as well as his letters and manuscripts, researchers are working toward linking these collections digitally to allow scholars to use them in answering new questions on biodiversity and colonial economics. This is a small part of what is becoming the digital humanities field. A massive effort centers on Charles Darwin; its core project is to publish his correspondence both in print and online, along with supporting materials.[13] Many of his plant specimens, which are held primarily at Cambridge University Herbarium, have been digitized, but not linked to the other Darwin materials. One problem in making this happen is that the science/humanities divide is still real in terms of funding and how data are handled.

The disconnect between science and the humanities that sometimes makes collaboration challenging seems to diminish when artists enter herbaria. In part this is because botanical art has always been essential to inquiry in the field. Before photography, illustrations were key in communicating knowledge about plant form. This remains true today, with most illustrations in scientific journals now in pen and ink. Drawings are frequently done from herbarium specimens, though artists do favor live material if available. When photography was developed, with its supposedly superior ability to present visual information objectively, it was assumed that botanical illustration would become obsolete, but this has hardly been the case. Photographs are not as good at providing clear depth cues, and artists can eliminate or underplay extraneous characteristics while clearly portraying the essential ones. Cameras and human vision work differently; the latter makes sense of the world by emphasizing boundaries and the outlines of objects, while cameras treat each piece of information the same. In other words, an illustrator does just what the visual system does. It's no wonder that humans find such images satisfying and understandable.[14]

While black-and-white printing might seem to militate against watercolor botanical art, the field has burgeoned in the last thirty years as an art movement. Periodic exhibits of recent work, such as those held at the Hunt Institute for Botanical Documentation in Pittsburgh and the Royal Horticultural Society, London, highlight this art, as have the efforts of Shirley Sherwood, a patron of the genre. She has amassed an impressive collection, publishing books and funding a gallery at Royal Botanic Gardens, Kew.[15] Botanic gardens have also discovered botanical illustration as a way to increase attendance, not only by holding exhibits but by establishing classes and certificate programs in botanical and natural history illustration. These generate interest in the field while adding practitioners with impressive skills and also enriching herbaria. Not only do botanical artists consult herbaria, many deposit specimens. Celia Rosser painted all known species in the iconic Australia genus *Banksia* and contributed vouchers from the plants she used as models to the herbarium of the Royal Botanic Gardens Victoria in Melbourne.[16]

At the same time that botanical art is seeing a resurgence, photography is an ever-growing influence in the botanical world, though there has long been a connection. With her seaweed cyanotypes, Anna Atkins was the first to illustrate a book with photographs, and another pioneer in the medium, amateur botanist Henry Fox Talbot, considered accurate specimen photographs a practical application of the technique. Now there are wildflower guides full of color photos as well as scientific journal articles with photographs to illustrate structural differences among species. Photos are sometimes attached to herbarium sheets to convey the plant's color, three-dimensional form, and scale. The botanist Arthur Cronquist often included field photos he took with his hat set near the plant as a size indicator; for consistency, it was always the same type of wide-brimmed straw hat.[17]

Digital photography and related software make it possible to create extremely clear images. Niki Simpson's painstaking photo collages, which resemble herbarium sheets, combine a photo of the intact "specimen" with various dissections, along with a notation system for structural information. Simpson is interested in botanical documentation, but in the process, she creates works of great beauty. Laurence Hill presents the genus *Fritillaria* in a website where he documents species with photographs of an entire plant and of each of its parts. Hill's more creative work speaks to the interplay between art and science. *Nature Diptych (Art)* is a large work with multiples of one *Fritillaria imperialis* image, reminiscent of Andy Warhol and representing similarity within a species, juxtaposed with photographs of specimens presenting species variability (fig. 16.3).[18] As another example

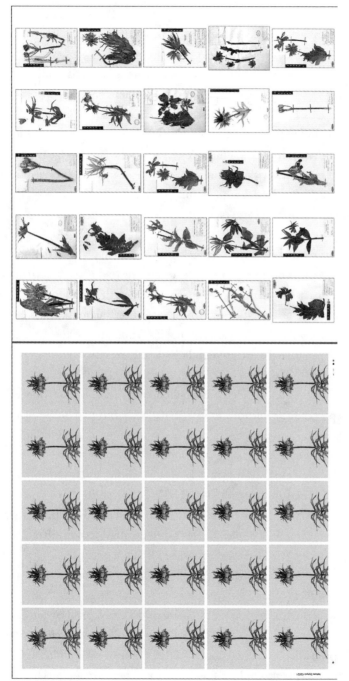

Figure 16.3. *Nature Diptych (Art)* (2021), by Laurence Hill, juxtaposes multiples of a photograph of a live plant of *Fritillaria imperialis* with a series of specimens of the same species. (Courtesy of Laurence Hill)

of specimens linked to popular culture, Rachel Pedder-Smith's massive work *Herbarium Specimen Painting* was used in fabric for a Vivienne Westwood designer dress.

Victoria Crowe is a well-known British artist who was commissioned to paint a portrait of David Ingram, noted botanist and former master of St. Catharine's College, Cambridge. Crowe's practice is to paint a background relating to the subject's interests. For Ingram this included an herbarium specimen. In preparation Crowe visited the Cambridge Herbarium and found much more than simply a specimen to sketch for her portrait. She became so intrigued that she returned to the herbarium and spent considerable time sketching and painting watercolors of specimens.

In a multimedia work, *Five Specimens Senecio paludosus, Fen Wort* (2007), she even drew the tape that held a specimen in place on the sheet, an almost unheard of detail in illustrating specimens (fig. 16.4). Crowe did not attempt to make the plants more lifelike; she recorded the wrinkled petals and faded colors just as they were on the sheets. She found that she was taken with specimens' combination of fragility and timelessness. Crowe then went further, studying early modern herbals and creating an exhibit entitled *Plant Memory*.[19] She conveyed a deep message about the passage of time with images from early printed herbals along with her specimen drawings.

As part of their outreach efforts, many herbaria open their collections to artists, inviting them to use specimens as inspiration and grounding for their work. One striking example is the Australian artist Gregory Pryor's *Black Solander* (2005) referencing Daniel Solander, the Swedish botanist who collected in Australia with Joseph Banks on Captain James Cook's first voyage there. Pryor worked in the Western Australian Herbarium and made small drawings of every one of the approximately 10,500 plant species known in the region. These were done in black ink on black sugar paper to suggest loss of biodiversity and the hidden toll of colonization on plants and indigenous people (fig. 16.5).[20]

The German artist Anselm Kiefer employs dried plants extensively in his work, though not in ways as closely tied to herbarium specimens. In his *Evil Flowers* (1985–92), dried sprays of delphiniums covered in shellac are pasted to an oil painting, almost obscuring it. These are not nicely pressed

Figure 16.4. A mixed-media work, *Five Specimens, Senecio paludosus Fen Wort*, by Victoria Crowe from her *Plant Memory* exhibition, 2007. The central image is drawn from an herbarium specimen and even includes the tape. (© Victoria Crowe)

Figure 16.5. Detail from Gregory Pryor's *Black Solander* (2005), one of the artist's 10,500 drawings in ink, pencil, and spirit-based inks on black sugar paper. (Installation at Perth Institute of Contemporary Arts, courtesy of the artist)

plants—they are brown, with flowers gone to seed—yet they are still tall and stately. The title suggests an ominous story, as does much of his work. For Kiefer, dried plants are more a matter of metaphor than science. He was born in Germany in March 1945 and as a child he considered it normal to play in piles of rubble. But Kiefer's work is not totally about destruction and death; it is also about memory and preservation. His sculpture *For Paul Celan—Ukraine* (2005) resembles an herbarium, a massive pile of lead sheets with aluminum sunflowers pressed between them, their flower heads and stems sticking out at each end. It is as if the plants are struggling to leave the confines of the sheets and find the sun. There is hope here, as in many of Kiefer's works that incorporate sunflowers: real dried plants, real seeds, and painted or sculpted plants. His art, complex both visually and conceptually, reveals how dried plant material can play a role in artistic expression as well as in botanical science.[21]

The Herbarium as Metaphor

Kiefer's use of dried plants as a metaphor both for decay and for renewed life suggests the herbarium's metaphorical power. In fact, the present use of *herbarium* to refer to plant collections arose from a metaphor. Long before plants were pressed, the word referred to a book on plants, often illustrated. Otto Brunfels's 1534 *Herbarum Vivae Eicones* was an early herbal, a volume on medicinal plants that had realistic illustrations. This was about the same time that pressed plants were first created, but the term used for such collections was usually the Latin *hortus siccus,* dry garden, or *hortus hyemalis,* winter garden. Agnes Arber writes: "The word 'Herbarium,' in the modern sense, makes its first appearance in print—so far as the present writer is aware—in Pitton de Tournefort's *Elemens* of 1694." Tournefort used a common term in a new way, somewhat metaphorically, not for a collection of plant illustrations but for a collection of the plants themselves. The nineteenth-century naturalist Sadie Price played with this connection by using herbarium sheets for her fern drawings and adding a specimen label, naming herself as collector, but with no specimens included.[22]

Those working in herbaria are often called on to describe their place of employment to family and friends. Say the word, and people automatically think of growing and selling herbs, or being into health food or herbal medicine. Their faces usually fall when they find that herbaria house dead plants, all labeled with scientific names. The Latin root of this term, *herb,* means herb or grass. It has been tied to plant material for millennia, and "herbarium" continues to be used in a variety of ways, usually to describe plant-related collections, as in the case of a book on the portrait artist Lucian Freud's paintings of plants, *Lucian Freud Herbarium.*[23] Such metaphorical usage can cause some confusion: Freud, for all his interest in plants and his attention to rendering them realistically, never pressed plants.

The philosopher Max Black argues that a metaphor is different from a direct comparison: "man is a wolf" is more powerful than "man is like a wolf." The principal subject, man, and the subsidiary subject, wolf, have similarities and differences, but it is the similarities that are heightened in a metaphor. In the case of herbarium, using the term "Lucian Freud herbarium" emphasizes that Freud painted a collection of plants as a way to preserve something of them visually, even though he did not press the plants or dry them or paste them onto sheets of paper. There have been several collections of plant-related art referred to as herbaria, including a British exhibit called simply *Herbarium*; a book on art linking technology with plants, *The Technological Herbarium*; and a plant photography album, *Suburban Herbarium*. The artwork *Herbarium Vivum* is composed of large flat boxes with each displayed vertically, like large herbarium sheets. Each box has a glass face and a plant growing in the narrow space between the glass and the white background, which includes a label, as herbarium specimens would. This is indeed a living herbarium, interrogating the relationship between living plant and specimen.[24]

Right now it's fashionable to link the herbarium metaphorically with the current interest in plants as sentient beings or at least worthy of more attention for having sophisticated interactions with the environment, including with other living things. It's no coincidence that Aloi, author of the Lucian Freud book, is also editor of a volume of essays, *Botanical Speculations,* on plants in contemporary art with a focus on them as active agents in the environment. There is also *Botanical Drift: Protagonists of the Invasive Herbarium,* a collection of conceptual art pieces presented at Kew, which encouraged visitors to appreciate plants in new ways.[25]

Michael Marder, a philosopher and leading figure in "critical plant studies" linking the humanities to botany, uses the herbarium metaphor in two different senses in his writing. In *The Philosopher's Plant: An Intellectual Herbarium,* he describes how a number of philosophers incorporated their observations on plants into building their philosophical systems, with Aristotle using the development of a wheat grain as an example of potentiality. Marder's *The Chernobyl Herbarium* is more personal. He describes living downwind from Chernobyl at the time of the reactor accident and includes photographs by Anaïs Tondeur. She pressed plants grown near the reactor onto photographic plates developed by radiation the cuttings emitted. The resulting images are ghostly but powerful, linking the herbarium to ideas of deterioration, death, and fragility.[26]

Max Black contends that metaphors employed too frequently lose their force, but "herbarium" is hardly there yet, though there is a line of fragrant soaps and a brand of gin each called Herbarium. Metaphorical uses of the word may be one way it can gain currency. Then there is the idea of turning the metaphor around, making "herbarium" the principal subject, as in "an herbarium is a fungal mycelium." Mycelia are the threadlike structures that can wrap around the roots of plants and even grow into plant cells. Sometimes they are parasitical but often they're beneficial to both parties, with fungi absorbing nutrients from the plant while also funneling soil minerals to it, and even moving nutrients and signaling molecules among plants in the area. Herbaria in the digital age are becoming more like this, linking their collections to those of other institutions, with all gaining from the interconnections. As fungi seem to act as hubs for communication among parts of the living subterranean world, specimens are serving as hubs for the proposed Extended Specimen Network. Just as fungi are hidden and underappreciated, so are herbaria, yet, also like fungi, they are coming into their own, with exciting new inquiry in the twenty-first century.

Herbaria can also be seen as hidden gardens. Like gardens, they contain selected plants that have been carefully prepared for inclusion. In both cases, it is obvious that humans have been involved and are necessary to their continued existence. On the other hand, living gardens require water while herbaria have to be kept away from it, and at least some people would consider living plants more beautiful than dead ones. Herbaria have definitely been a well-kept secret until now, but renewed interest in nature is making them bloom more publicly.

Herbaria Blooming

THOUGH IT MAY NOT SEEM THAT WAY, I've only skimmed the surface of herbaria's rich past, present, and future. I could have written much more—on the techniques and experience of making herbarium specimens: to attach with linen tape, paste, thread, or (horrors) Scotch tape; to attempt to dry specimens in steamy rainforests; to strive for artful placement of a specimen on a sheet. And there are thousands of collectors and millions of specimens left unmentioned, along with numberless adventures and misadventures. But my aim here has been less to be comprehensive and more to open a door on this realm of hidden gardens. Living gardens contribute so much to the happiness and well-being of many, and biodiversity is essential to continuance of life on this planet. Hidden gardens are crucial to both these more obvious worlds in documenting their past and present and planning for their future.

Herbaria must become less secreted in order to foster greater appreciation for the living world and the record we have created of it. Science, history, art, and the experience of being human will be enriched by such a prospect. This bold statement stems from the breadth of herbaria's influences. First and foremost, herbaria and other natural history collections are the bedrock of biological inquiry. They are not just documents about life on Earth but physical evidence of it. Besides recording the diversity of life and the dynamics of ecosystems, they reveal much about genetics, morphology, and evolution. The information they contain can guide future efforts to conserve that diversity and therefore improve the quality of life on Earth.

Collections are also references in the search for new ways to use this diversity as medicines, foods, fibers, and so much more.

The obvious necessity of herbaria to biological science sometimes masks their value as historical reference material, not only for the history of science but, as has been revealed in many ways here, for the history of exploration and colonization, and for social history. Herbaria are being mined for information on the role of women, indigenous and enslaved persons, and others whom the written botanical record has slighted, but who have contributed substantially to that record. Horticulture, science, and the humanities are so intertwined that herbarium specimens are important links in creating a fuller view of who was involved in creating the gardens of the past and present. Art historians are discovering that herbaria can reveal much about how artists of the past used dried as well as living plants in their work. Herbarium collections and individual specimens can be treated as works of art as well as of science in terms of how the plants were arranged and embellished. This aesthetic aspect is also pivotal to the work of artists now studying plant collections and using them as inspiration for contemporary work.

These different enterprises need to be linked because they are connected even within single herbarium sheets. Some sheets from the George Clifford collection can be type specimens because Carl Linnaeus consulted them in describing species. Many of these plants came from Dutch colonies in areas in Africa and the Far East where the Dutch East India Company, of which Clifford was a director, held sway, creating wealth by exploiting these outposts. Its ships brought back specimens, bulbs, and plants that were then cultivated and preserved by Clifford and his circle of botanical allies. So his specimens are tied not only to Linnaeus but to colonial history, commerce, and horticulture. Finally, there is the aesthetic appeal of the sheets; this was important to Clifford since he considered them as status symbols, with their elaborate labels and printed vase embellishments.

In present-day Madagascar, botanists from former colonial nations participate in long-term projects of plant collecting coupled with building a firm infrastructure for future research and conservation. Specimens serve as records of the many plants that grow on the island and nowhere else; dozens of new species are identified each year. The saved material often includes samples specially preserved so they can be used in DNA sequencing. Specimens also document indigenous names and uses as well as the changing relationships among botanists, as Madagascan professionals are frequently now the collectors of record, and they may be assisted by parataxonomists in their newly professionalized role. A complete set of specimens remains in the country; these are often tied to projects in developing medicines as well as sustainable agriculture and forestry. Here biodiversity, ethnobotany, economics, and sociology can be intertwined in one specimen.

Herbarium sheets document the human experience of plants in all its richness. The quality of all our lives is diminished if they are kept hidden. Today's curators are finding novel and exciting new ways to open their collections physically and virtually to new constituencies while also forging stronger links with the broader scientific community, ecologists and geneticists as well as molecular and evolutionary biologists. Stronger, more robust bridges to all those interested in plants are necessary if herbaria are to become better known and more ably utilized to answer new questions and to revisit old ones. The first herbaria were created to preserve information about medicinal plants and also novel species; these collections were designed to assist in studying plants and to teaching others about them. These uses of "dry gardens" are still relevant, but now there are many more ways they feed human curiosity. They will do so even more in the future as they become less shrouded.

Notes

Introduction

1. The curator was the late Marilyn Massaro, who was very gracious and helpful to me on subsequent visits to the Roger Williams Park Museum of Natural History.
2. Spamer and McCourt, "The Lewis and Clark Herbarium," tracing the travels of the Lewis and Clark specimens, 10–26.
3. Harkness, *The Jewel House,* 31.
4. Curtis, "The Philosopher's Flowers"; Morgenroth, "Research from Home," on linking Martin's herbarium sheets to his drawings; Cage, *A Mycological Foray;* Luxemburg, *Herbarium.*
5. Cook, *Jean-Jacques Rousseau,* describes the working collection of Abraham Gagnebin, 128.
6. Armstrong, "John Bartram and Peter Collinson," the working relationship between Bartram and Collinson, 23–37; Fara, *Sex, Botany, and Empire,* deals with Banks's unofficial role in directing British colonial botanical enterprises, 47–95.
7. Specimen #124981 at the A. C. Moore Herbarium, University of South Carolina, Columbia, collected on October 24, 2018, by John Nelson, then the herbarium's curator.
8. Holton, *Thematic Origins of Scientific Thought,* describes the private side of science, 19–24; Evert and Eichhorn, *Raven Biology of Plants,* botany textbook; Gribbin and Gribbin, *Flower Hunters.*
9. Damasio, *The Feeling of What Happens,* a book-length argument for the relationship between cognition and affect; Dewey, *Art as Experience,* on true experience as having an aesthetic component, 35–58; Adamo et al., "Plant Scientists' Research Attention," findings on the influence of morphological and color traits in collecting, 574.
10. Yang, "That Drunken Conversation between Two Cultures," uncertainty as fundamental in both art and science, 319–20; Root-Bernstein, *Discovering,* discussion of scientists as artists, 312–40.
11. Wilson and Peter, *Biodiversity,* the first collection of papers that helped to popularize the term *biodiversity;* Wilson, *Biophilia,* 3; Leopold, *A Sand County Almanac,* on the conservation aesthetic, 155–56.

12. McCarthy, *The Moth Snowstorm,* the power of aesthetics in driving environmentalism, 157–69; Mackenzie et al., "We Do Not Want to 'Cure Plant Blindness,' " moving plants to the foreground, 139.

Chapter 1. Rooted in an Herbarium

1. Angell and Romero, "Orchid Illustrations," Oakes and Blanche Ames in Berlin, 20.
2. Plimpton, *Oakes Ames,* contains many excerpts from letters between Blanche and Oakes Ames, most written by the latter; Ames, *Orchidaceae,* systematic treatment of the orchid family in seven volumes.
3. Clark, *My Dear Mrs. Ames,* quotes 71–73.
4. Garay, "The Orchid Herbarium of Oakes Ames," describes Ames's donations to the Harvard University Herbarium, 49.
5. Balick, "Reflections on Richard Evans Schultes," outlines Schultes's life and work, 7, 8–10.
6. Plimpton, *Oakes Ames,* quote 75.
7. Ames and Ames, *Orchids at Christmas;* Ames and Ames, *Drawings of Florida Orchids.*
8. Brown, Fulton, and Pfister, *Glass Flowers of Harvard,* on the history of the collection; Ames, *The Ware Collection,* Oakes Ames's publication on the glass flowers.

Chapter 2. Early Botany

1. The term *botanist* did not arise until the eighteenth century, but I will use it here to designate those interested in natural history with a focus on plants. Baldini, Cristofolini, and Aedo, "The Extant Herbaria," Ghini's plant collecting, 23; Toresella and Battini, "Gli erbari," early examples of pressed leaves, 72–73.
2. Ogilvie, *The Science of Describing,* classification of the generations of early modern botanists, 28; Fahnestock, "Forming Plants in Words and Images," sources of medical knowledge, 4; Findlen, "The Death of a Naturalist," Ghini's methods of learning about plants, 129.
3. Blunt and Stearn, *The Art of Botanical Illustration,* on early printed herbals, 61–71; Brunfels, *Herbarum Vivae Eicones;* Fuchs, *De Historia Stirpium.*
4. Arber, *Herbals,* gives the history of herbaria, 138–43.
5. Findlen, "The Death of a Naturalist," Ghini's move from Bologna to Pisa, 133–36.
6. Bellorini, *The World of Plants in Renaissance Tuscany,* describes the Medicis' relationship with botany, 40–44.

7. Stefanaki et al., "Breaking the Silence," about Petrollini's herbarium in Leiden, 3.

8. Egmond, "Into the Wild," describes early modern botanical field trips, 176.

9. Latour, "Drawing Things Together," on immutable mobiles, 6; Müller-Wille and Scharf, "Indexing Nature," allowing facts to travel, 2.

10. Findlen, "The Death of a Naturalist," relationship between Ghini and Aldrovandi, 136–42.

11. Findlen, *Possessing Nature,* Aldrovandi's collections, 346–47.

12. Ciancio, "The Many Gardens of Mattioli," Mattioli's contributions to botany, 36.

13. Findlen, "The Death of a Naturalist," *Placiti* was published after Ghini's death, 146; Camus, "Historique des premiers herbiers," Ghini sending specimens and notes to Mattioli, 296.

14. Mattioli, *Senensis Medici.*

15. Star and Griesemer, "Institutional Ecology," on boundary objects, 388.

16. Morton, *History of Botanical Science,* describes Cesalpino's approach to classification, 128–41; Bellorini, *The World of Plants in Renaissance Tuscany,* Cesalpino's work in the context of new plant discoveries, 70–76.

17. Egmond, *Eye for Detail,* Gessner's working method, 66, 126; Blair, "Dedication Strategies," Gessner's dedications in his publications, 170–71.

18. Gessner, *Historiae Animalium;* Kusukawa, *Picturing the Book of Nature,* Gessner's note taking, 139–61; Zoller, Steinmann, and Schmid, *Conradi Gesneri Historia Plantarum* is a German edition of Gessner's two botanical notebooks at the library of the University of Erlangen-Nuremberg.

19. Kusukawa, "Drawing as an Instrument of Knowledge," on *Nicotiana,* 41–42.

20. Dewey, *Art as Experience,* on the elements of an aesthetic experience, 35–58; Bleichmar, *Visible Empire,* describes techniques involving the senses used in learning about plants, 65.

21. Egmond, *The World of Carolus Clusius,* gives an excellent insight into Clusius's botanical travels and writings.

22. Ommen, *The Exotic World of Carolus Clusius,* translations of others' works by Clusius, 10.

23. Pavord, *The Naming of Names,* Clusius in Vienna, 58–64.

24. Clusius, *Rariorum Plantarum Historia,* was published in 1601 and *Exoticorum Libri Decem* in 1605.

25. Egmond, *The World of Carolus Clusius,* on Clusius's correspondence, 11.

26. Harkness, *The Jewel House,* discusses the Lime Street naturalists, including their links to Clusius, 15–56.

Chapter 3. The Technology and Art of Herbaria

1. Basbanes, *On Paper,* the introduction of paper into Europe, 57–58.

2. Hunter, *Papermaking,* review of the history of paper and its production, 77–169.

3. Egmond, *Eye for Detail,* Gessner's use of notes and sketches, 66; Daston and Galison, *Objectivity,* on four-eyed sight and its importance in scientific illustration, 84–97.

4. Kusukawa, *Picturing the Book of Nature,* the history of Gessner's illustrations, 139.

5. Morton, *History of Botanical Science,* Pliny's description of an illustrated herbal by Crateuas (120–60 BCE), 66; Harris, *Roots to Seeds,* John Sibthorp studied the drawings copied from the *Juliana Codex* in Vienna before his collecting trip to Greece, 94; Ivins, *Prints and Visual Communication,* image copying before printing, 14–16.

6. Blunt and Stearn, *The Art of Botanical Illustration,* plant illustrations in medieval manuscripts, 32–36; Pächt, "Early Italian Nature Studies," describes the Carrara Herbal created around 1400, 27.

7. Smith, *The Body of the Artisan,* observation of nature as important to development of art, 95.

8. Arber, *The Mind and the Eye,* quote on art as a way to communicate information about plant form, 21.

9. Daniel et al., "*Anisotes* in Madagascar," an artist's contribution to correcting a plant description, 121; Stevenson and Stevenson, "The Nuts and Bolts," another example of a botanical artist's keen eye, 142; Nasim, "James Nasmyth's Lunar Photography," drawing as learning, 171.

10. Arber, *Herbals,* describes the history of the early printed herbals, 13–78; Kusukawa, *Picturing the Book of Nature,* part 2 provides a good look into how printed herbals with illustrations were created in the sixteenth century.

11. Brunfels, *Herbarum Vivae Eicones.*

12. Fuchs, *De Historia Stirpium;* Saunders, *Picturing Plants,* differences in plant depictions between Fuchs and Brunfels, 9–12.

13. Egmond, *Eye for Detail,* describes the importance of collections of naturalia drawings to understanding early modern natural history illustrations, 6–13.

14. Ibid., plants depicted in isolation, 100; Gross and Harmon, *Science from Sight to Insight,* images as theory-laden, 72.

15. Elliott, Guerrini, and Pegler, *Flora,* Cesi's use of a microscope, 26.

16. Koning et al., *Drawn After Nature,* the relationship of Clusius to the *Libri Picturati,* 13.

17. Cave, *Impressions of Nature,* a good guide to the history of nature prints; Tognoni, "Nature Described," Colonna's nature prints and engravings, 361–68.
18. Olariu, "The Misfortune of de Lignamine's Herbal," on the Fuchs print with leaf additions, 55.
19. Benkert, "The 'Hortus Siccus,'" Platter's enhancements to specimens, 231–34.
20. Ibid., Platter's use of Weiditz watercolors, 228–29.
21. Harder, *Historia Stirpium,* Oak Spring Garden Library, Upperville, VA, Record 439; Kusukawa, "Image, Text and '*Observatio,*'" use of herbaria by apothecaries, 468; Dobras, "Hieronymus Harder," the elder Harder's herbaria, 46–82.
22. Harris, *Planting Paradise,* in the 1690s, the London nurseryman William Darby used an herbarium as a marketing tool, 34.
23. Thijsse, "Herbaria of George Clifford III," on elaborate labels, 136; Prince, *Stuffing Birds, Pressing Plants,* on Haüy's coloring method, 43.
24. Mabey, *The Cabaret of Plants,* quote, 27.

Chapter 4. Early Exploration

1. Cooper, "Placing Plants on Paper," surveying forests of Saxe-Gotha, 257.
2. Thinard, *Explorers' Botanical Notebook,* where and what Belon collected, 40; Duval, *The King's Garden,* Belon acclimatizing exotic species, 14.
3. Andel, "Open the Treasure Room and Decolonize the Garden," recent study of the Rauwolf herbarium, 8; Ghorbani et al., "Historical Herbarium of Leonhard Rauwolf," relationship between Rauwolf specimens and illustrations, 568.
4. Duval, *The King's Garden,* Tournefort fearful of shipwrecks, 51.
5. Oviedo, *Historia general de las Indias.*
6. Bleichmar, *Visual Voyages,* description of pineapple, 21, and chocolate, 59–63.
7. Steele, *Flowers for the King,* information from indigenous physicians, 5; materials sent to Spain, 6
8. Čermáková and Černá, "Naked in the Old and New World," on species not mentioned by ancient authors, 70; Appleby, *Shores of Knowledge,* Hernández's work, 104.
9. Hernández et al., *Nova Plantarum.*
10. Bleichmar, *Visual Voyages,* Monardes's approach to studying American plants, 50–57; Monardes, *De Simplicibus Medicamentis;* Monardes, *Joyfull Newes out of the Newfound Worlde.*
11. Appleby, *Shores of Knowledge,* about the natural history survey of New Holland, 110; Piso and Marcgraf, *Historia Naturalis Brasiliae.*

12. Andel et al., "The Forgotten Hermann Herbarium," discussion of the significance of an herbarium from Hendrik Meyer in Surinam, attributed to Paul
Hermann, 1296.

13. Egmond, *The World of Carolus Clusius,* Canadian seeds being grown in France
and distributed widely, 126.

14. Ibid., network of botanists sharing seeds and specimens, 128.

15. Potter, *Strange Blooms,* describes Tradescant's North American plant collection, 265; Wulf, *The Brother Gardeners,* on Tradescant, 10.

16. Kastner, *A Species of Eternity,* on Banister, 10.

17. Sargent, "Recentering Centers of Calculation," the complexities of medical
practice in colonies, 298.

18. Ibid., description of van Rheede's project; 303–10; van Rheede, *Hortus Indicus
Malabaricus.*

19. Winterbottom, "Medicine and Botany in the Making of Madras," a case study
on two physician/botanists in India in the seventeenth century, 35–57; Latour,
"Drawing Things Together," centers of calculation, 59; Sargent, "Recentering
Centers of Calculation," on van Rheede's work in India, 302–12.

20. Fraser and Fraser, *The Smallest Kingdom,* Hermann's collections in South
Africa, 37.

21. Andel and Barth, "Paul Hermann's Ceylon Herbarium," on Hermann's travels, 977–78; Ubrizsy Savoia, "The Influence of New World Species," geographic origin of plants mistaken in early modern botany, 166.

22. Stearn, introduction to *Species Plantarum,* Carl Linnaeus borrowed these
specimens, giving them scientific interest since many are described and named
in his *Flora Zeylanica,* 119.

23. Harbsmeier, "Fieldwork Avant La Lettre," Kaempfer's botanical work in Japan,
34–35.

24. Crane, *Ginkgo,* on Kaempfer and gingko, 195; Kaempfer, *History of Japan* and
Amoenitatum Exoticarum on Japanese natural history; Nakamura, *Kinmo zui,*
a massive illustrated natural history of Japan.

25. Dandy, *The Sloane Herbarium,* Hermann's plants in the Sloane Herbarium, 145.

26. George, *William Dampier in New Holland,* Dampier's collecting on the Western Australian coast, 1.

27. Madriñán, *Nikolaus Joseph Jacquin's American Plants,* on Jacquin's need to
document plants in drawings, 5; Lack, "The Plant Self Impressions Prepared
by Humboldt and Bonpland," 220.

28. Fraser and Fraser, *The Smallest Kingdom,* Thunberg collecting in South Africa, 62; Corner, *Botanical Monkeys,* describes how he trained monkeys to collect for him.

Chapter 5. The Value of Collecting

1. Findlen, *Possessing Nature,* Aldrovandi's collections and publications, 7–30.
2. Delbourgo, *Collecting the World,* a good review of Sloane's life and collections.
3. Ibid., the intellectual foundations of the Royal Society, 28–30.
4. Stungo, "Specimens from the Chelsea Physic Garden," a history of the Society of Apothecaries specimens sent to the Royal Society each year, 213.
5. Sloane, *Natural History of Jamaica;* Delbourgo, *Collecting the World,* Kick's blending of information from watercolors and specimens in his work, 107; Rose, "Natural History Collections and the Book," juxtaposition of Jamaican plants and drawings of them, 16.
6. Delbourgo, *Collecting the World,* quote on Sloane as collector, 202; MacGregor, *Curiosity and Enlightenment,* Bobart's advice to Sloane on collecting, 130.
7. Riley, "The Club at the Temple Coffee House," describes the membership and activities of the group.
8. Stearns, "James Petiver," how Petiver cultivated relationships with collectors, 287–310.
9. Ibid., quote from Petiver's instructions on collecting, 365; Fleischer, "Leaves on the Loose," describes the extent of Petiver's collections and their organization, 124–26.
10. Murphy, "James Petiver's 'Kind Friends,' " the importance of the slave trade to Petiver's collecting, 259; Delbourgo, *Collecting the World,* Sloane and slavery in Jamaica, 71–86.
11. Delbourgo, *Collecting the World,* significant contribution of Sloane's labels and catalogues, 259; Blom, *To Have and to Hold,* the functions of catalogues in collecting, 215.
12. Delbourgo, *Collecting the World,* Ray's *Historia Plantarum* as reference for the Sloane Herbarium, 159; Rose, "Natural History Collections and the Book," Sloane's *Natural History of Jamaica* as a guide to Jamaica plants in his herbarium, 15–33.
13. Dandy, *The Sloane Herbarium,* Linnaeus's comment on Sloane's herbarium after his visit to London, 8.
14. Pickering, *Putting Nature into a Box,* an analysis of the extant portion of Sloane's Vegetable Substances collection.
15. Goff, "The Schildbach Wood Library," describes the origin and history of this collection; Avis-Riordan, "How Kew's Wood Experts Solve Crimes."
16. Young, "Brendel Plant Model Survey"; Tribe, "The Dillon-Weston Glass Models," models of microfungi, 169.

17. Thijsse, "Herbaria of George Clifford III," elaborate labels and decorative additions on specimens, 140–44; Tomasi and Willis, *An Oak Spring Herbaria*, the history and characteristics of Carlo Sembertini's herbarium, 334–39.

18. Stuckey, "The First Public Auction of an American Herbarium," the fate of various portions of Zacheus Collins's herbarium, 447–54.

19. Kohlstedt, "Museum Perceptions and Productions," quote from Mason, 23.

20. Shepherd, Thiedemann, and Lehnebach, "Genetic Identification of Historic *Sophora*," tracing the history of a *Sophora toromiro* specimen, 7.

21. Carine, *The Collectors*, discusses the stories behind the specimens in Sloane's herbarium.

22. Das and Lowe, "Nature Read in Black and White," decolonial approaches to interpreting and presenting natural history collections.

Chapter 6. Linnaeus and Classification

1. Freer, *Linnaeus's Philosophia Botanica*, 18.

2. Blunt, *The Compleat Naturalist*, Linnaeus's student years in Uppsala, 24–37.

3. Linnaeus, *Flora Lapponica*.

4. Ray, *Historia Plantarum;* Tournefort, *Elemens de botanique*.

5. Morton, *History of Botanical Science*, Joachim Jung presented clear descriptions of many plant structures, 167–74; Vaillant developed the idea of male and female reproductive organs within a flower, 241–42.

6. Stearn, introduction to *Species Plantarum*, how straightforward Linnaean classification was, 17–24.

7. Merriman, "Peter Artedi," Linnaeus's work with Artedi and the latter's death, 34–38.

8. Linnaeus, *Systema Naturae;* Rutgers, "Linnaeus in the Netherlands," visit to Gronovius, 107.

9. Blunt, *The Compleat Naturalist*, the geographic range of the plants in each of Clifford's hothouses, 100–101.

10. Ibid., Boerhaave quote, 110.

11. Ibid., William Houston's collection, 112.

12. Calman, *Ehret*, on interactions between Linnaeus and Ehret at Hartekamp, 49–50.

13. Linnaeus, *Hortus Cliffortianus* and *Musa Cliffortiana*, which described the Clifford banana plant.

14. Linnaeus, *Systema Naturae, Fundamenta Botanica, Classes Plantarum, Genera Plantarum*.

15. Juel and Harshberger, "New Light on the Collection," Kalm's activities during his North American trip, 297–300; Robbins, "Jane Colden," Kalm's visit to the Colden's, 46; Colden, *Botanic Manuscript.*

16. Fara, *Sex, Botany, and Empire,* Linnaeus's former students sought and found many new species, but none were of real economic benefit in Sweden, 34–37.

17. Linnaeus, *Öländska och Gothländska Resa.*

18. Linnaeus, *Philosophia Botanica* and *Species Plantarum;* Jarvis, *Order out of Chaos,* provides an analysis of the importance of *Species Plantarum* in botanical nomenclature, 1–60.

19. Gardiner and Morris, *The Linnaean Collections,* significance of *Species Plantarum,* 4; Baack, "A Naturalist of the Northern Enlightenment," Forsskål's use of Arabic plant names, 9; Blunt, *The Compleat Naturalist,* Thunberg's travels and writings, 194–97.

20. White, "The Purchase of Knowledge," describes Smith's acquisition of the Linnaean collections and its outcomes, 126–29; Müller-Wille, "Linnaeus' Herbarium Cabinet," three cabinets were sent to London: one remains at the Linnean Society, London, while the other two were returned to Uppsala in 1938, 60.

21. Freer, *Linnaeus's "Philosophia Botanica,"* on Linnaeus's elements for classification, 129.

Chapter 7. Botanical Exploration

1. Bleichmar, *Visible Empire,* describes the botanical impetus behind the late eighteenth-century expeditions sponsored by the Spanish government, 3–15.

2. Bleichmar, "The Geography of Observation," the importance of art to Mutis's project, 383.

3. Blunt and Stearn, *The Art of Botanical Illustration,* on the Sessé and Mociño illustrations and specimens, 92; Bleichmar, *Visible Empire,* on Monciño and de Candolle, 70–72.

4. McVaugh, *Botanical Results of the Sessé and Mociño Expedition.*

5. Dickenson, *Drawn from Life,* natural history laboratories on naval ships, 213.

6. Williams, *Naturalists at Sea,* Commerson's background, 58; Duval, *The King's Garden,* Poivre's work in Mauritius, 82; Williams, *French Botany in the Enlightenment,* Commerson's collections, 9–11.

7. Williams, *Naturalists at Sea,* Labillardière's collection, 213–17.

8. Scurr, *Napoleon,* botanical work in Egypt, 50, and André Thouin went to Italy specifically to select specimens and also exotic plants for the Jardin des plantes in Paris, 83; Figueiredo and Smith, "Joaquim José da Silva," from the Royal

Botanical Garden of Madrid, the French naturalist Geoffroy St. Hilaire brought to Paris choice specimens, including some collected in Brazil and Angola, 131.

9. Williams, *French Botany in the Enlightenment,* importance of notes and specimens in allowing Labillardière to describe the species he collected, 174–76; Mulvaney, *The Axe Had Never Sounded,* the extent of his collections, 83; Webb, *The Botanical Endeavor,* Webb purchased the collection at auction in 1834, 50.

10. O'Brian, *Joseph Banks,* Banks's work with Solander over years of botanizing and exploration, 61–146.

11. Ewan and Ewan, *John Banister and His Natural History of Virginia,* the relationship between care at the time of collection and the value of the final specimen, 173.

12. Smith, *European Vision and the South Pacific,* the many erasures and revisions in Solander's notebooks attest to the difficulties he had in classifying Australian plants, 166; Banks, Solander, and Parkinson, *Banks' Florilegium.*

13. Fara, *Sex, Botany, and Empire,* extent of Miller's herbarium, 135, and George III, Banks, and Kew, 131; Goodman, *Planting the World,* deals with the plant collectors that Banks sent out from Kew.

14. Aiton et al., *Hortus Kewensis;* Goodman, *Planting the World,* describes the collectors, their discoveries, and travel difficulties.

15. Helferich, *Humboldt's Cosmos,* Humboldt quoted on the variety of organisms they encountered, 190; Adamo et al., "Plant Scientists' Research Attention," a study on collection bias toward larger plants with showier flowers, 574.

16. Wulf, *The Invention of Nature,* preparations for the expedition, 39–48.

17. Ibid., Humboldt and Bonpland travel herbarium, 76.

18. Lack, "The Botanical Field Notes," the botanical notebook entries, 503.

19. Helferich, *Humboldt's Cosmos,* travels in Mexico and the United States, 265–99.

20. Humboldt, *Essai sur la géographie des plantes.*

21. Lack, *Alexander von Humboldt,* learning from Forster, 15.

22. Anthony, "Mining as the Working World of Alexander von Humboldt," how Humboldt's years as a mining engineer influenced his interest in rock strata and plant variation in different types of rock and soil, 1.

23. Moret et al., "Humboldt's *Tableau Physique* Revisited," the relationship between the plants Humboldt and Bonpland collected and recent collections from the same area; quote, 12889.

24. Lack, *Alexander von Humboldt,* Kunth's work in Paris, 63, and his retrieving the botanical notebooks, 74; Granados-Tochoy, Knapp, and Orozco, "*Solanum humboltianum,*" on a newly described species first collected by Humboldt and Bonpland, 200.

25. Humboldt, *Cosmos;* Arber, "Goethe's Botany," translation of Goethe's *The Metamorphosis of Plants* has an introduction to Goethe's ideas on plant morphology and scientific inquiry; Bonpland, Humboldt, and Kunth, *Nova Genera et Species Plantarum.*

Chapter 8. Gardens

1. Read, *Art and the Evolution of Man,* the adaptive advantage of an aesthetic sense, 12.

2. McKibben, *The End of Nature,* the first two chapters lay out his argument on human disturbance of every part of the Earth, 3–80.

3. Vines, "Robert Morison," influence of his time working in France, 16–23.

4. Arber, *Herbals,* Pepys seeing Evelyn's herbarium, 142.

5. Pickering, *Putting Nature into a Box,* breadth of Somerset's botanical and horticultural interests and knowledge, 233.

6. Davies, "Botanizing at Badminton," Somerset's melancholy and how gardening and botany helped her to overcome it, 24–26.

7. McClain, *Beaufort,* describes shipments from Barbados, 19.

8. Davies, "Botanizing at Badminton," Sherard and Beaufort's working relationship, 26.

9. Ibid., Sloane asked Somerset to grow plants for a Royal College of Physicians project, 33.

10. Cottesloe and Hunt, *The Duchess of Beaufort's Flowers,* on a Somerset South African import, 39; McClain, *Beaufort,* Somerset's use of Kick's drawings for her embroidery, 214; LaBouff, "Embroidery and Information Management," study of the embroidery of Mary, Queen of Scots and Bess of Hardwick in empowering women intellectually.

11. Cook, *Jean-Jacques Rousseau,* plant study as focusing the mind, 15; Rousseau's interest in native plants, 21.

12. Ibid., botanical letters to Madeleine Delessert, 190; Stafleu, "Benjamin Delessert," Delessert's massive herbarium and its organization, 921–33.

13. Laird, *A Natural History of English Gardening,* description of the botanical activities of Bentinck and Delany in this and the following paragraphs drawn largely from 167–92.

14. Leapman, *The Ingenious Mr. Fairchild,* description of Fairchild's mule, 167–68.

15. Jeppe, *Herbarium Vivum,* catalogue with specimens of German grasses; Tomasi, *An Oak Spring Flora,* Furber's floral calendar catalogue, 143–46.

16. Catesby, *The Natural History of Carolina, Florida and the Bahama Islands,* in volume 1 text accompanying plate 27 on dogwoods, Catesby mentions that they grew at Thomas Fairchild's nursery, where Catesby worked for a time.

17. Wulf, *The Brother Gardeners,* how Collinson and Bartram developed their partnership, 27–33.

18. Note on specimen of *Prunus pensylvanicus,* page 6 of volume 12, Lord Petre Herbarium, Sutro Library, San Francisco.

19. O'Neill and McLean, *Peter Collinson,* what Bartram supplied to Petre, 66–68; Ewan, "Plant Collectors in America," Petre's botanical advisers, 32.

20. Schuyler and Newbold, "Vascular Plants in Lord Petre's Herbarium," the Bartram specimens in the Sutro Library, 42.

21. Gunn and Codd, *Botanical Exploration,* Oldenland's work in South Africa and his herbarium, 265.

22. Bleichmar, *Visible Empire,* investigations on cinnamon in South America, 134–36; Stearns, *Science in the British Colonies,* varying views on Chinese and American ginseng, 288–89.

23. Crawford, *The Andean Wonder Drug,* difficulties in determining quinine concentrations in *Cinchona* and how to chemically stabilize it, 1–4.

24. Brockway, "Science and Colonial Expansion," plant transfers and colonial powers, 456–58.

25. Wallich, *Plantae Asiaticae Rariores;* Sivasundaram, "The Oils of Empire," Wallich's collecting in Burma, 386.

26. Noltie, *Indian Forester,* environmental problems in India due to deforestation, 81–89.

27. Ibid., Hugh Cleghorn's work for the East India Company evaluating forestry, 81–104; Brockway, *Science and Colonial Expansion,* describes the role of British colonial gardens in transplanting species and developing plantations.

Chapter 9. Managing Exploration and Collecting

1. Lawrence, *Adanson,* Adanson's work in Senegal and in publishing species descriptions on his return, 20–40; Linnaeus's naming *Adansonia,* 41.

2. Stoffelen et al., "Central African Plants," about the Central African plant collections in Belgium.

3. Bynum and Bynum, *Botanical Sketchbooks,* Kirk's collecting on Livingstone's second Zambezi expedition, 27; Polhill and Polhill, *East African Plant Collectors,* the origins of the flora of tropical East Africa project, 5–7.

4. Erickson, *The Drummonds of Hawthornden,* Drummond's difficulties in collecting and his relationship with Preiss, 36–44; Haebich, "The Forgotten German Botanist," deals with Preiss's Australian collecting.

5. Goodman, *Planting the World,* Banks's role in choosing Brown and Bauer for the Flinder's expedition, 245; Hewson, *Australia,* Bauer's drawings and plates of Australian plants, 40–49.

6. Noltie, *John Hope,* review of Hope's contributions, 8–13; Noltie, *Botanical Art from India,* provides a good treatment of the work of Indian botanical artists.

7. Latour, "Drawing Things Together," centers of calculation, 59; Teltscher, *Palace of Palms,* discussion of the work of William Jackson Hooker and Joseph Dalton Hooker at Kew, 72–93.

8. Endersby, *Imperial Nature,* analysis of different approaches to classification and difficulties between collectors and taxonomists, 137–69.

9. Webb, *The Botanical Endeavor,* the interactions between Mueller and Bentham, 226–33; Bentham and Mueller, *Flora Australiensis.*

10. Watt, *Robert Fortune,* argument for why Fortune's methods in obtaining tea plants were not underhanded, 267–69.

11. Bishop, *Travels in Imperial China,* deals with David's expeditions, including the discovery of *Davidia,* 95.

12. Musgrave, Gardner, and Musgrave, *The Plant Hunters,* Wilson's connection with Augustine Henry in finding *Davidia,* 156–58.

13. Kastner, *A Species of Eternity,* Lewis receiving instruction from Benjamin Barton, 128; Barton, *Elements of Botany.*

14. McKelvey, *Botanical Explorations,* Pursh's questionable behavior, 74.

15. Spamer and McCourt, "The Lewis and Clark Herbarium," traces the travels of the Lewis and Clark Expedition specimens, 7–14.

16. Evelyn, "The National Gallery at the Patent Office," on the disposition of the Wilkes Expedition collection in Washington, DC, 232.

17. Eyde, "Expedition Botany," negotiations between Wilkes and Gray, 38.

18. Dupree, *Asa Gray,* Gray received the Walter specimen from Lindley, 80; Dampier specimen is #BM001041517 in the herbarium at the Natural History Museum, London.

19. Darlington, *Flora Cestrica.*

20. John Torrey to William Darlington, January 18, 1853: original manuscript in the William Darlington Papers, New York Historical Society Mss Collection, vol. 8, item 125.

21. Torrey, *On the Darlingtonia Californica.*

22. William Darlington to John Torrey, February 2, 1853: original manuscript in the John Torrey Papers, New York Botanical Garden, Torrey PP, series 1, folder

2.44, letter to Torrey in which Darlington described Bentham as "inexorable"; Flannery, "Naming a Genus for William Darlington," on the *Darlingtonia* story.

23. Ewan, *Rocky Mountain Naturalists,* collecting careers of Charles Parry, 278–79, and William Emory, 203–4.

24. Birch, "A Comparative Analysis of Nineteenth Century Pharmacopoeias," on Lincecum's learning from indigenous medical practitioners, 430.

25. Ravenel, *Fungi Caroliniani Exsiccati;* Childs, *Journal of Henry William Ravenel,* receiving financial assistance, 318, support in selling books and collections, 398, on trip to Texas, 332.

26. Mauz, *C. G. Pringle,* how Sargent worked with Pringle, 28; Marsh, *Man and Nature,* a work on human intervention in nature that influenced future generations of environmentalists.

27. Mauz, *C. G. Pringle,* quote from Engelmann's letter, 32.

Chapter 10. Natural History and Botany

1. Barber, *The Heyday of Natural History,* provides an introduction explaining why natural history was so popular in the nineteenth century and its links to natural theology, 13–26.

2. Pearce, "John Stuart Mill's Botanical Collections," Mill's collecting, particularly later in life, 152–53.

3. Holway, *The Flower of Empire,* Paxton's work for the Duke of Devonshire on glasshouses, 92–104.

4. Phelps, *Familiar Lectures on Botany;* Dickinson, "Dickinson/Abiah Root (Strong) Correspondence," quote from letter to Abiah Root dated May 7, 1845; Dickinson, *Emily Dickinson's Herbarium,* presents a facsimile of the poet's specimen collection with a commentary on its composition and conservation.

5. Stearn, "Mrs. Agnes Arber," on how silent work helped her thinking, 262; Doty, *Still Life with Oysters and Lemon,* on describing in words and images, 6.

6. Secord, "Corresponding Interests," working-class collectors and their botanical interactions with those in other classes, 383–93.

7. Cribb and Tibbs, *A Very Victorian Passion,* information on Day's collection and drawings, 12–38; relationship with Reichenbach, 7.

8. Hayward and Rickard, *Fern Albums,* chapter on New Zealand's fern albums, 53–140.

9. DiNoto and Winter, *The Pressed Plant,* souvenir herbaria were popular during the Victorian era, 28.

10. Atkins, *Photographs of British Algae;* Bryant et al., "Margaret Gatty," Gatty's work and her collaboration with botanist William Henry Harvey, 138–39.

11. Kniphof, *Herbarium Vivum,* and Ludwig, *Ectypa Vegetabilium,* both illustrated with nature prints; Cave, *Impressions of Nature,* on Auer's method and his dispute with Bradbury, 94–105; Ettingshausen and Pokorny, *Physiotypia Plantarum Austriacarum,* highlights Auer's approach; Moore, *The Ferns of Great Britain and Ireland;* Bradbury, Johnstone, and Croall, *The Nature-Printed British Sea-weeds.*

12. Teltscher, *Palace of Palms,* ownership transfer of the elder Hooker's herbarium to Kew, 298.

13. Ibid., a commission recommended the rehabilitation of Kew, 19–47.

14. MacGregor, *Company Curiosities,* East India Company plant collections transferred to Kew, 187.

15. Endersby, *Imperial Nature,* dispute over the Kew and NHM herbaria, 296–99.

16. Dupree, *Asa Gray,* hiring the next generation of Harvard botanists, 351–53.

17. Mickulas, *Britton's Botanical Empire,* review of NYBG's early growth, 209–14.

18. Daniel, "One Hundred and Fifty Years of Botany at the California Academy of Sciences," on Alice Eastwood's role in saving CAS collections after the San Francisco earthquake, 241–42.

19. Barber, *The Heyday of Natural History,* argument that natural history became "dull" once it became a school subject, 294.

20. Kohler, *All Creatures,* extent of U.S. survey work, 10–15.

21. Ibid., Bessey's studies in Nebraska, 104.

22. Quinn, *Windows on Nature,* introduction to how and why dioramas are created, 8–23; producing plant material for exhibits, 153–57.

23. Nesbitt and Cornish, "Seeds of Industry and Empire," Kew's economic botany collections, 62–64; Diagre-Vanderpelen, "The Rise and Fall of the Belgian Forestry Museum."

24. Hetherington and Dolan, "Roots in the Paleobotanical Collection," microscope slide of root tip suggesting early diversity in root structure, 4.

Chapter 11. Evolution and Botany

1. Darwin, *The Botanic Garden,* Erasmus Darwin's poetic works on botany.

2. Kohn et al., "What Henslow Taught Darwin," Henslow's views on plant variation, 643.

3. Ibid., Henslow's influence on Darwin's intellectual development, 645.

4. Allan, *Darwin and His Flowers,* Henslow's reaction to Darwin's specimens, 84.

5. Desmond, *Sir Joseph Dalton Hooker*, Hooker's analysis of the plants Darwin collected on the Galápagos Islands, 254.

6. Malthus, *Essay on the Principle of Population;* Ziman, *Public Knowledge,* in chapter 5, defines public knowledge as that published or presented in a scientific forum.

7. Darwin, *A Monograph of the Subclass Cirripedia.*

8. Charles Darwin to Asa Gray, April 17, 1855, Asa Gray Papers, Harvard University Botanical Library, Cambridge, MA.

9. Gray, "Diagnostic Characters," describes Japanese plants.

10. Linnaeus, *Species Plantarum.*

11. Stevens, *The Development of Biological Systematics,* how the natural system developed through the work of Antoine de Jussieu and others.

12. de Candolle and de Candolle, *Prodromus Systematis Naturalis Regni Vegetabilis;* de Candolle, *Lois de la nomenclature botanique.*

13. Nicolson, "A History of Botanical Nomenclature," review of efforts to stabilize botanical names.

14. Kingsland, *The Evolution of American Ecology,* controversy over the botanical code, 43–59.

15. Sloane, *Natural History of Jamaica;* Turland et al., *International Code.*

16. Daston, "Type Specimens," exposition on the type specimen concept.

17. Hildreth et al., "Standard Operating Procedure for the Collection and Preparation of Voucher Plant Specimens," vouchering commercial batches of medicinal plant preparations, 16; Culley, "Why Vouchers Matter," specimens as vouchers for plant identification using DNA sequencing, 1–2.

18. Manzano and Julier, "How FAIR Are Plant Sciences?" field studies are not reproducible without vouchers, 20202597; Beatrix Farrand's herbarium is at the University of California, Berkeley.

19. Turland et al., *International Code,* chapter 4 on what qualifies as effective publication in systematics.

20. Bebber et al., "Herbaria Are a Major Frontier for Species Discovery," new species found among herbarium specimens, 22169; Lack and Mabberley, *The Flora Graeca Story,* a campanula from Crete described by Paolo Boccone from a plant growing in an Italian botanical garden, 7.

21. Le Bras et al., "The French Muséum national d'histoire naturelle," number of unmounted specimens in the Paris herbarium, 4; Borsch et al., "A Complete Digitization of German Herbaria," unmounted specimens in German herbaria, 8; Próćków et al., "How Many Type Specimens?" type specimens discovered in Polish collections that had been moved many times.

22. Flora of China Project, *Flora of China;* Editorial Committee, *Flora of North America;* Polhill and Polhill, *East African Plant Collectors,* history of the Flora of East Africa project, 5; participants, 6; Grace et al., "Botanical Monography," the publication of a flora leads to more robust taxonomic work, 433.

23. Cronquist et al., *Intermountain Flora.*

24. Holmgren, "History of the Intermountain Flora Project," provides background on the botanists involved in collecting plants and writing plant descriptions for the flora, 60–70.

Chapter 12. Changing Botany

1. Whitfield, "Superstars of Botany," Thomas Croat's collecting efforts, 437; Mori, "My Career as a Tropical Botanist," "baling hay" as a term for collecting specimens, 4.

2. Morton, *History of Botanical Science,* background on plant morphology and physiology, along with taxonomy, 232–62.

3. Cittadino, *Nature as the Laboratory,* the German botanical landscape in the nineteenth century and how it influenced the development of ecology, 3–9.

4. Kingsland, *The Evolution of American Ecology,* Clements's ideas on succession and how others responded to them, 142–57.

5. Brown, "Catesby in Carolina," describes Catesby's observations of damage to forests in clearing land for agriculture, 11.

6. "From a Garden in Brno," most of Mendel's herbarium was preserved in his monastery, but a few specimens were given to other institutions, 91; Leapman, *The Ingenious Mr. Fairchild,* the first artificial plant hybridization, 1.

7. Harman, *The Man Who Invented the Chromosome,* ill feeling between Darlington and herbarium curator, 223.

8. Eigsti, "Induced Polyploidy," cause and effects of polyploidy, 153–55.

9. Judson, *The Eighth Day of Creation,* introduction to the discovery of DNA's function and structure, 9–195.

10. Nevill et al., "Large-Scale Genome Skimming," herbarium specimens as source of DNA for sequencing, 1–2.

11. Crawford, "Plant Macromolecular Systematics," how sequencing data became important in plant systematics.

12. Andel et al., "Sixteenth-Century Tomatoes in Europe," recent work on the En Tibi herbarium in the Naturalis Biodiversity Center in Leiden, 1–2; Attwood et al., "Concepts, Historical Milestones and the Central Place of Bioinformatics," basics of bioinformatics, 3.

13. Hagen, "Five Kingdoms," discovery of fundamental differences among fungi, many algae, and plants, 67–68.

14. Turland et al., *International Code;* Chase and Reveal, "APG III"; Chase et al., "APG IV."

15. Kull and Rangan, "Science, Sentiment and Territorial Chauvinism," botanical and political issues that came to the fore in the *Acacia* nomenclatural controversy; Robin and Carruthers, "National Identity and International Science," historical and cultural significance of the *Acacia* debate, 49.

16. Prather et al., "The Decline of Plant Collecting," data on acquisitions over a forty-year period in U.S. herbaria; Prather et al., "Implications of the Decline in Plant Collecting," why decreased specimen collecting is detrimental to future biodiversity research.

17. Winston, *Describing Species,* quote on 179.

18. Masson, "Brief History of the Collector's Index to the Wabash College Herbarium," provenance of the Custer specimens, 212–14.

19. Timby, "The Dudley Herbarium," merger of the Stanford and CAS herbaria, 3–5.

20. Ibid., David Akerly's reasons for needing a larger herbarium for his work, 13.

21. Holstein, "Had an Enquiry for a Single Cyperus Specimen."

22. Mori, "From the Field," Barneby's yearly plant identifications, 142.

Chapter 13. Useful Plants and Ethnobotany

1. Nesbitt, "Use of Herbarium Specimens," useful plants are more often collected, 313.

2. Strocchia, *Forgotten Healers,* women apothecaries and their practices in the early modern era, 4–11.

3. Sanjad, Pataca, and dos Santos, "Knowledge and Circulation of Plants," example of the influence of indigenous expertise from the Amazon region in the eighteenth century.

4. Balick and Cox, *Plants, People, and Culture,* nutmeg could be sold by the Dutch in Amsterdam for thirty thousand times what growers were paid, 136.

5. Fara, *Sex, Botany, and Empire,* Banks's role in managing the transport of breadfruit to the West Indies, 138.

6. Nesbitt and Cornish, "Seeds of Industry and Empire," the changing fortunes of economic botany collections, 62–63.

7. Cornish and Nesbitt, "Vegetable Sheep," objects in economic botany collections no longer have economic importance; they are cultural objects, 21; Milliken, "Mobilising Richard Spruce's 19th Century Amazon Legacy," Kew's

collaborative efforts with Brazilian botanical institutions using materials collected in Brazil by Spruce.

8. Andel, "Open the Treasure Room," botanical collection at the Naturalis Biodiversity Center in the Netherlands; Das and Lowe, "Nature Read in Black and White," review of issues in decolonizing collections.

9. Murphy, "James Petiver's 'Kind Friends,' " slave trade used to facilitate plant collecting.

10. Mueggler, *The Paper Road*, the complex relationships among Forrest and his collectors, especially with his chief assistant, Zhao Chengzhang, 46–60.

11. Sheldrake, "The 'Enigma' of Richard Schultes," basic focus of ethnobotany, 347.

12. Balick, "Reflections on Richard Evans Schultes."

13. *Convention on International Trade in Endangered Species.*

14. *Convention on Biological Diversity;* Crawford, *The Andean Wonder Drug,* provides a history of the use of cinchona in South America and beyond.

15. *The Nagoya Protocol.*

16. Novotny and Molem, "An Inventory of Plants," difference between the two halves of New Guinea in the ease of doing biological research, 533.

17. Blumberg, "Case Study of Plant-Derived Drug Research," description of an herbarium related to medical research, 6–9.

18. Balick and Cox, *Plants, People, and Culture* is an overview of ethnobotany drawn from the experiences of two outstanding researchers in the field.

19. Osseo-Asare, *Bitter Roots,* herbarium specimens in her research, 23; Voeks, *The Ethnobotany of Eden,* the jungle medicine narrative, 4–5.

20. Balick and Cox, *Plants, People, and Culture,* differences between uses of medicines in developed and developing countries, and the criteria to evaluate them, 56–59.

21. Eisenman, Tucker, and Struwe, "Voucher Specimens," vouchers document medicinal plants in manufacturing of herbal medicines, 30.

22. Nabhan, *Where Our Food Comes From,* Vavilov's theory of geographic centers where cultivated crops were developed, 29–64.

23. Mahapatra, "Return of Indigenous Crops," tribal groups in the Indian state of Odisha are using indigenous crops to improve food security.

24. Nabhan, *Ethnobiology for the Future,* papers on links between preserving indigenous cultures and environments; Dold and Cocks, "Preliminary List of Xhosa Plant Names," indigenous plant names recorded on specimens collected for a cancer study of Bantu people, 267–68.

25. Basu, "N. W. Thomas Botanical Collections," study of an anthropologist's plant specimens; Basset et al., "Conservation and Biological Monitoring," training parataxonomists in the tropics, 163.

26. Cámara-Leret, Fortuna, and Bascompte, "Indigenous Knowledge Networks," loss of indigenous languages leads to loss of plant knowledge, 9914–15; Nabhan and St. Antoine, "The Loss of Floral and Faunal Story," experience of plants and plant knowledge reduced among younger generations, 239–45.

27. Mucioki, "Creating Herbaria," tribal herbarium as part of a food security project; Thiers, *Index Herbariorum,* information on the Tebiwa Herbarium of the Newe people.

Chapter 14. Understanding and Conserving Biodiversity

1. Wilson and Peter, *Biodiversity,* first collection of essays to popularize the term *biodiversity.*

2. Antonelli et al., *State of the World's Plants and Fungi, 2020*; Nesbitt and Cornish, "Seeds of Industry and Culture," history of economic botany collections at Kew, 54–63; Thiers, *Index Herbariorum,* tracks herbaria based on size as well as other criteria, 7.

3. Willis, *State of the World's Fungi,* fungal characteristics and evolution, 6–17.

4. Heberling and Burke, "Utilizing Herbarium Specimens to Quantify Historical Mycorrhizal Communities," describes studies of fungal growth on specimens, 1–2; Heberling and Isaac, "Herbarium Specimens as Exadaptations," using research techniques that had not been developed when the specimens were collected, 963–65; Parker, Schanen, and Renner, "Viable Soil Algae," identification of algae from soil taken on specimens; Meineke et al., "Herbarium Specimens Reveal Increasing Herbivory," measuring insect damage on plant specimens, 105; Whitehead, "Collecting Beetles," an argument for hunting beetles by examining herbarium specimens, 249; Miquel and Bungartz, "Snails Found among Herbarium Specimens," micro-mollusks found on plant specimens, 173.

5. Roullier et al., "Historical Collections," comparing specimens from early explorers and later botanists, 2206.

6. Cámara-Leret et al., "New Guinea Has the World's Richest Island Flora," report on a recent survey of New Guinea plant diversity, 579–80.

7. Colli-Silva et al., "Evaluating Shortfalls," collecting biases in herbarium specimens from the Atlantic Forest region of Brazil, 567–70; Cowell, Anderson, and Annecke, "Historic Herbarium Specimens," collecting on the Agulhas Plain in South Africa, 483–84; Demissew et al., "Sub-Saharan Botanical Collections," results of plant collecting in the Cameroon, 106.

8. Heberling, Prather, and Tonsor, "The Changing Uses of Herbarium Data," findings on herbarium use, 132; Funk, "100 Uses for an Herbarium."

9. Kuzmina et al., "Using Herbarium-Derived DNAs," project to barcode Canadian plants.

10. Gutaker et al., "The Origins and Adaptation of European Potatoes," European potatoes related to Andes landraces, 1093; Goss et al., "The Irish Potato Famine Pathogen *Phytophthora infestans*," Mexican sources for the pathogen that caused the Irish potato famine, 8791; Campos et al., "Deciphering the Invasive History of a Bacterial Crop Pathogen"; Dieuliis, Bajema, and Winstead, "Biospecimens and the Information Landscape for Biodefense," why biological specimen collections are important in bioterrorism defense.

11. Large et al., "Changes in Stomatal Density over Time," relationship between stomata number and atmospheric CO_2 levels, 319–20.

12. Moore and Lauenroth, "Differential Effects of Temperature and Precipitation," temperature affected spring-flowering species while precipitation influenced fall-flowering species.

13. Kudo and Cooper, "When Spring Ephemerals Fail to Meet Pollinators," relationship between timing of bee emergences and orchid pollination, 1; Václavík et al., "Effects of UV-B Radiation," lack of evidence for a link between UV radiation and leaf hairs, e0175671.

14. Zangerl and Berenbaum, "Increase in Toxicity," herbarium specimens contained higher amounts of toxic substances after the arrival of an invasive insect species, 15529; Ent et al., "X-Ray Fluorescence," plant species with high nickel levels in several tissues, 15–16.

15. *Convention on Biological Diversity*; *Global Strategy for Plant Conservation*, the updated version of the original plan; *World Flora Online* and its ancillary site, *Online Floras*; Brazil Flora Group, "Brazilian Flora 2020," describes the project to create a comprehensive online flora of algae, fungi, and plants.

16. Silvertown, *Demons in Eden*, bias against rare plants in Red List, 139.

17. Ahlstrand and Stevenson, "Retracing Origins of Exceptional Cycads," conservation and propagation of rare cycads at botanic gardens, 94–97.

18. Thiers, *The World's Herbaria, 2020*, report based on data from *Index Herbariorum;* the index provides data on the size and noteworthy collections in each institution, designated by an acronym, such as K for Kew and NY for New York Botanical Garden. This is a useful shorthand for referring to herbaria.

19. Avis-Riordan, "How Kew's Wood Experts Solve Crimes," Kew's xylarium provides evidence for law cases involving illegal timber trade and other crimes.

20. Wolkis et al., "Germination of Seeds," viable seeds from herbarium specimens, e576.

21. Eastwood et al., "Conservation Roles," comparison of the world's two major seed banks.
22. "Ross Potato Herbarium"; "Hidden Treasures," collections in the UC Davis herbarium, 27.
23. Mains, "Joseph Charles Arthur," contributions that Arthur's collection allowed him to make to the field, 602–4.
24. Lavoie, "Should We Care about Purple Loosestrife?" how the plant entered and spread in North America, 1967–68.
25. Casas et al., "Ethnobotany for Sustainable Ecosystem Management," analyzes sustainable agriculture programs in several areas of Mexico; Miller, *Plant Kin,* multispecies food production among the Canela indigenous people in Brazil, 63–76.

Chapter 15. Online Herbaria

1. Funk, "Collections-Based Science," review of digitization efforts and the value of natural history collections.
2. Shetler, "The Herbarium: Past, Present and Future," statistics on herbaria, 699; Lanjouw, *Index Herbariorum.*
3. Shetler, *Introduction to the Botanical Type Specimen Register,* digitizing information about type specimens; Russell, "A Brave New Digital World," problems arising from early adoption of technology, 3; Irwin et al., "America's Systematics Collections," plans for digitizing collections, 30.
4. Park et al., "The Colonial Legacy of Herbaria," uneven global distribution of specimens; Smith, "The African Plants Initiative," description of the project.
5. *Global Plants.*
6. Strasser, "The Experimenter's Museum," the similarities between management of DNA sequence collections and natural history collections, 61–64; Colella et al., "The Open-Specimen Movement," 1–3; *Global Biodiversity Information Facility.*
7. *Biodiversity Heritage Library*; *BHL Flickr.*
8. Sassoon, "Photographic Meaning," issues relevant to digitizing plant specimens, including about materiality and context, 310–12; Flannery, "Flatter Than a Pancake," what is gained and lost in digital imaging of herbarium specimens, 228–29; *International Image Interoperability Framework (IIIF),* software to present digital images in a standardized format across institutions.
9. "A Strategic Plan for Establishing a Network," report that led to the Advancing Digitization of Biodiversity Collections project; *iDigBio.*
10. *SERNEC*; *SEINet.*

11. Le Bras et al., "The French Muséum national d'histoire naturelle," renovation of the Paris herbarium and digitization of its collection, 2–5; Marsico et al., "Small Herbaria," unique contributions of small herbaria to documenting species diversity, 1577.

12. Boyle et al., "The Taxonomic Name Resolution Service," reasons for a name resolution service, 2.

13. Chandra, *Geek Sublime,* why software is difficult to update without creating unforeseen problems, 111–18.

14. *Encyclopedia of Life*; *World Flora Online*; "Online Floras," floras that are online but not in a format compatible for upload to *World Flora Online.*

15. *Canadensys*; *Environmental Information Reference Center*; *Atlas of Living Australia*; *China's National Specimen Information Infrastructure*; Europe's *Distributed System of Scientific Collections.*

16. Hedrick et al., "Digitization and the Future," 5–7.

17. Thiers et al., "Extending U.S. Biodiversity Collections," illustration of the extended specimen network, 24; Anderson, *Plants, Man and Life,* quote, 47; how Anderson developed his "inclusive herbarium" concept, 220–23.

18. *An Alliance for Biodiversity Knowledge.*

19. Manzano and Julier, "How FAIR Are Plant Sciences?" characteristics of FAIR data, 20202597.

20. Heberling, "Herbaria as Big Data Sources," on ecologists' past lack of interest in herbarium specimens, 90; proposal to add new types of information to herbarium sheets, 108–9.

21. Feeley and Silman, "The Data Void in Modeling," number of specimens per species needed in modeling studies, 626.

22. Yost et al., "The California Phenology Collections Network," project to determine phenological status of digitized herbarium specimens, 132–33.

23. Pearson et al., "Machine Learning," extracting phenological data from images, 610–11.

24. Lendemer et al., "The Extended Specimen Network," need for continued collecting, 26–27.

Chapter 16. A Broader Vision

1. McCarthy, *Moth Snowstorm,* relationship between appreciation of nature's beauty and efforts to preserve it, 245; Cowan, "The Herbarium as a Databank," herbaria store information, 3; Macmillan, *Victoria Crowe,* the herbarium as a botanical memory bank, 20.

2. Thiers, *Herbarium;* LaFarge, *Herbaria.* The Society of Herbarium Curators was founded in 2004.

3. *Biodiversity Literacy in Undergraduate Education.*
4. Barber, *The Heyday of Natural History.*
5. Drinkwater, "Festival Fun with Frankenstein's Plants," an herbarium activity as part of the Royal Botanical Garden's festival.
6. Hester, "Preserve Your Quarantine Nature."
7. Li, Pryer, and Windham, "*Gaga*, a New Fern Genus"; Martine et al., "*Solanum Watneyi*"; Nagy, "Reborn Herbarium."
8. Megan, "UConn Finds Two Plant Specimens"; Young, "Brendel Plant Model Survey."
9. Spellman and Mulder, "Validating Herbarium-Based Phenology," citizen science in the herbarium, 898.
10. *Zooniverse*; "Notes from Nature"; Hill et al., "The Notes from Nature Tool," how the Notes app works and users' responses to it, 217–29.
11. Hughes, "Digital Collections as Research Infrastructure," transcription projects increase appreciation for archives; Gianquitto, " 'My Dear Dr.,' " frontier women supplying information and specimens to Asa Gray and other botanists, 362–70; Endersby, *Imperial Nature*, Joseph Hooker's correspondence with amateur botanists in Australia and New Zealand, particularly William Colenso, 105–7, and Ronald Gunn, 149–50.
12. Heberling et al, "Data Integration," decline in the percentage of specimens in the GBIF database, 2; Heberling and Isaac, "iNaturalist as a Tool," linking observations with specimens; *iNaturalist*; *Leafsnap.*
13. "Natural History Museum to Collaborate"; *Darwin Correspondence Project.*
14. Kingdon, "In the Eye of the Beholder," analysis of differences between how eye and camera function, 137–39.
15. Among Sherwood's books are *Contemporary Botanical Artists* and, with Harris and Juniper, *A New Flowering.*
16. Rosser and George, *The Banksias;* Charles, *Highgrove Florilegium;* Charles, Akeroyd, and Mills, *The Transylvania Florilegium.*
17. Thomas, *Beauty of Another Order,* Atkin's cyanotypes, 92, and Talbot's plants, 78; Weiss, "Art Cronquist's Hat," gives examples of Cronquist's specimens with photographs attached.
18. Simpson and Barnes, "Photography and Contemporary Botanical Illustration," on creating composite images, 265–69; Hill, *Fritillaria Icones,* photographs and information on all the species in the genus.
19. Crowe and Ingram, *Plant Memory,* passage about fragility and timelessness, 9.
20. Aloi, "Gregory Pryor," analysis of *Black Solander,* 35.

21. Biro, *Anselm Kiefer,* discussion of dried plants and memory, 10, and sunflowers, 71, 86; Batsaki, "The Apocalyptic Herbarium," analysis of Kiefer's use of plant material to explore the biodiversity crisis, 401.

22. Arber, *Herbals,* on Tournefort's definition of an herbarium, 142; Flannery, "Sadie Price."

23. Aloi, *Lucian Freud Herbarium.*

24. Black, "Metaphor," 286–88; Gatti, *The Technological Herbarium;* Arnold, *Suburban Herbarium;* Driessens and Verstappen, *Herbarium Vivum.*

25. Aloi, *Botanical Speculations,* on turning away from an anthropocentric view of plants, xxxiii; Carroll, *Botanical Drift.*

26. Marder, *The Philosopher's Plant,* Aristotle and a grain of wheat, 32; Marder, *The Chernobyl Herbarium.*

Bibliography

Abe, Naoko. *The Sakura Obsession: The Incredible Story of the Plant Hunter Who Saved Japan's Cherry Blossoms.* New York: Knopf, 2019.

Adamo, Martino, Matteo Chialva, Jacopo Calevo, Filippo Bertoni, Kingsley Dixon, and Stefano Mammola. "Plant Scientists' Research Attention Is Skewed towards Colourful, Conspicuous and Broadly Distributed Flowers." *Nature Plants* 7, no. 5 (2021): 574–78. https://doi.org/10.1038/s41477-021-00912-2.

Ahlstrand, Natalie Iwanycki, and Dennis W. Stevenson. "Retracing Origins of Exceptional Cycads in Botanical Collections to Increase Conservation Value." *Plants, People, Planet* 3, no. 2 (2021): 94–98. https://doi.org/10.1002/ppp3.10176.

Aiton, William, Franz Andreas Bauer, Georg Dionysius Ehret, George Nicol, and James Sowerby. *Hortus Kewensis.* London: Nicol, 1789. https://www.biodiversitylibrary.org/item/23432.

Allan, Mea. *Darwin and His Flowers: The Key to Natural Selection.* New York: Taplinger, 1977.

An Alliance for Biodiversity Knowledge. Accessed January 29, 2022. https://www.biodiversityinformatics.org/.

Aloi, Giovanni, ed. *Botanical Speculations.* Newcastle upon Tyne, UK: Cambridge Scholars, 2018.

———. "Gregory Pryor: Postcolonial Botany." *Antennae,* no. 18 (2011): 24–36.

———. *Lucian Freud Herbarium.* Munich: Prestel, 2019.

Ames, Blanche, and Oakes Ames. *Drawings of Florida Orchids.* Cambridge, MA: Botanical Museum of Harvard University, 1947.

———. *Orchids at Christmas.* 1975. Reprint, Cambridge, MA: Botanical Museum of Harvard University, 2007.

Ames, Oakes. *Orchidaceae: Illustrations and Studies of the Family Orchidaceae.* 7 vols. Boston: Houghton, Mifflin, 1905–22. https://doi.org/10.5962/bhl.title.15433.

———. *The Ware Collection of Blaschka Glass Models of Plants in the Botanical Museum of Harvard University.* Cambridge, MA: Botanical Museum of Harvard University, 1946.

Andel, Tinde van. "Open the Treasure Room and Decolonize the Museum." Inaugural lecture, Leiden University, 2017. https://www.universiteitleiden.nl/en/

research/research-output/science/inaugural-lecture-open-the-treasure-room-and-decolonize-the-museum.

Andel, Tinde van, and Nadine Barth. "Paul Hermann's Ceylon Herbarium (1672–1679) at Leiden, the Netherlands." *Taxon* 67, no. 5 (2018): 977–88. https://doi.org/10.12705/675.8.

Andel, Tinde van, Sarina Veldman, Paul Maas, Gerard Thijsse, and Marcel Eurlings. "The Forgotten Hermann Herbarium: A 17th Century Collection of Useful Plants from Suriname." *Taxon* 61, no. 6 (2012): 1296–1304. https://www.jstor.org/stable/24389114.

Andel, Tinde van, Rutger A. Vos, Ewout Michels, and Anastasia Stefanaki. "Sixteenth-Century Tomatoes in Europe: Who Saw Them, What They Looked Like, and Where They Came From." *PeerJ* 10 (2022): e12790. https://doi.org/10.7717/peerj.12790.

Anderson, Edgar. *Plants, Man and Life.* Berkeley: University of California Press, 1952.

Angell, Bobbi, and Gustavo A. Romero. "Orchid Illustrations at Harvard." *Botanical Artist* 17, no. 1 (2011): 20–21. https://www.jstor.org/stable/45037498.

Anthony, Patrick. "Mining as the Working World of Alexander von Humboldt's Plant Geography and Vertical Cartography." *Isis* 109, no. 1 (2018): 28–55. https://doi.org/10.1086/697061.

Antonelli, A., C. Fry, R. J. Smith, M. S. J. Simmonds, P. J. Kersey, H. W. Pritchard, M. S. Abbo, et al. *State of the World's Plants and Fungi, 2020.* Kew, UK: Royal Botanic Gardens, Kew, 2020. https://doi.org/10.34885/172.

Appleby, Joyce. *Shores of Knowledge: New World Discoveries and the Scientific Imagination.* New York: Norton, 2013.

Arber, Agnes. "Goethe's Botany." *Chronica Botanica* 10 (1946): 63–126.

———. *Herbals: Their Origin and Evolution, a Chapter in the History of Botany, 1470–1670.* 2nd ed. Cambridge: Cambridge University Press, 1938.

———. *The Mind and the Eye: A Study of the Biologist's Standpoint.* Cambridge: Cambridge University Press, 1954.

Armstrong, Alan W. "John Bartram and Peter Collinson: A Correspondence of Science and Friendship." In *America's Curious Botanist: A Tercentennial Reappraisal of John Bartram,* edited by Nancy E. Hoffmann and John C. Van Horne, 23–42. Philadelphia: American Philosophical Society, 2004.

Arnold, William. *Suburban Herbarium: Photography and Culture.* Axminster, UK: Uniformbooks, 2020.

Asia Biodiversity Conservation and Database Network: ABCDNet. Accessed January 30, 2022. http://www.abcdn.org/.

Atkins, Anna. *Photographs of British Algae: Cyanotype Impressions.* 5 vols. Seven-
    oaks, UK, 1843–53. https://digitalcollections.nypl.org/collections/photo
    graphs-of-british-algae-cyanotype-impressions#/?tab=navigation.
Atlas of Living Australia. Accessed January 29, 2022. https://www.ala.org.au/.
Attwood, T. K., A. Gisel, N.-E. Eriksson, and E. Bongcam-Rudloff. "Concepts,
    Historical Milestones and the Central Place of Bioinformatics in Modern
    Biology: A European Perspective." In *Bioinformatics—Trends and Method-
    ologies,* edited by M. A. Madhavi, 3–38. London: IntechOpen, 2011.
Avis-Riordan, Katie. "How Kew's Wood Experts Solve Crimes." *Watch and Read*
    (blog), May 9, 2019. Royal Botanical Gardens, Kew. https://www.kew.org/
    read-and-watch/wood-anatomy-science-solve-crime.
Baack, L. J. "A Naturalist of the Northern Enlightenment: Peter Forsskål After 250
    Years." *Archives of Natural History* 40, no. 1 (2013): 1–19. https://doi.
    org/10.3366/anh.2013.0132.
Baldini, Riccardo M., Giovanni Cristofolini, and Carlos Aedo. 2022. "The Extant
    Herbaria from the Sixteenth Century: A Synopsis." *Webbia* 77, no. 1 (2022):
    23–33. https://doi.org/10.36253/jopt-13038.
Balick, Michael. "Reflections on Richard Evans Schultes, the Society for Economic
    Botany, and the Trajectory of Ethnobotanical Research." In *Medicinal
    Plants and the Legacy of Richard E. Schultes,* edited by Bruce E. Ponman
    and Rainer W. Bussmann, 3–11. St. Louis: Missouri Botanical Garden,
    2012.
Balick, Michael J., and Paul Alan Cox. *Plants, People, and Culture: The Science of
    Ethnobotany.* New York: Scientific American, 1996.
Banks, Joseph, Daniel Solander, and Sydney Parkinson. *Banks' Florilegium.* 34
    vols. London: British Museum, 1981–88.
Barber, Lynn. *The Heyday of Natural History, 1820–1870.* Garden City, NY: Dou-
    bleday, 1980.
Barton, Benjamin Smith. *Elements of Botany.* Philadelphia, 1803. https://babel.
    hathitrust.org/cgi/pt?id=hvd.32044107229783&view=1up&seq=5&skin=
    2021.
Basbanes, Nicholas A. *On Paper: The Everything of Its Two-Thousand-Year His-
    tory.* New York: Knopf, 2013.
Basset, Yves, Vojtech Novotny, Scott E. Miller, George D. Weiblen, Olivier Missa,
    and Alan J. A. Stewart. "Conservation and Biological Monitoring of Tropical
    Forests: The Role of Parataxonomists." *Journal of Applied Ecology* 41, no. 1
    (2004): 163–74. https://doi.org/10.1111/j.1365-2664.2004.00878.x.
Basu, Paul. "N. W. Thomas Botanical Collections." *[Re:]Entanglements* (blog),
    October 26, 2020. https://re-entanglements.net/botanical-collections/.

Batsaki, Yota. "The Apocalyptic Herbarium: Mourning and Transformation in Anselm Kiefer's Secret of the Ferns." *Environmental Humanities* 13, no. 2 (2021): 391–413. https://doi.org/10.1215/22011919-9320211.

Bebber, D. P., M. A. Carine, J. R. I. Wood, A. H. Wortley, D. J. Harris, G. T. Prance, G. Davidse, et al. "Herbaria Are a Major Frontier for Species Discovery." *Proceedings of the National Academy of Sciences* 107, no. 51 (2010): 22169–71. https://doi.org/10.1073/pnas.1011841108.

Bellorini, Cristina. *The World of Plants in Renaissance Tuscany.* Farnham, UK: Ashgate, 2016.

Benkert, Davina. "The 'Hortus Siccus' as a Focal Point: Knowledge, Environment, and Image in Felix Platter's and Caspar Bauhin's Herbaria." In *Sites of Mediation,* edited by Susanna Burghartz, Lucas Burkart, and Christine Göttler, 211–39. Leiden: Brill, 2016. https://doi.org/10.1163/9789004325760_010.

Bentham, George, and Ferdinand von Mueller. *Flora Australiensis.* 7 vols. London: Reeve, 1863–78. https://www.biodiversitylibrary.org/item/3669.

BHL Flickr. Flickr. Accessed January 29, 2022. https://flickr.com/biodivlibrary.

"BiCIKL Homepage." Accessed November 15, 2021. https://bicikl-project.eu/.

Biodiversity Heritage Library. Accessed January 29, 2022. http://biodiversitylibrary.org/.

Biodiversity Literacy in Undergraduate Education (BLUE). Accessed January 29, 2022. https://www.biodiversityliteracy.com.

Birch, Joanne L. "A Comparative Analysis of Nineteenth Century Pharmacopoeias in the Southern United States: A Case Study Based on the Gideon Lincecum Herbarium." *Economic Botany* 63, no. 4 (2009): 427–40. https://doi.org/10.1007/s12231-009-9101-8.

Biro, Matthew. *Anselm Kiefer.* New York: Phaidon, 2013.

Bishop, George. *Travels in Imperial China: The Explorations and Discoveries of Père David.* London: Cassell, 1990.

Black, Max. "Metaphor." *Proceedings of the Aristotelian Society* 55 (1954): 273–94. https://doi.org/10.1093/aristotelian/55.1.273.

Blair, Ann. "The Dedication Strategies of Conrad Gessner." In *Professors, Physicians and Practices in the History of Medicine: Essays in Honor of Nancy Siraisi,* edited by Gideon Manning and Cynthia Klestinec, 169–209. New York: Springer, 2017.

Bleichmar, Daniela. "The Geography of Observation: Distance and Visibility in Eighteenth-Century Botanical Travel." In *Histories of Scientific Observation,* edited by Lorraine Daston and Elizabeth Lunbeck, 373–95. Chicago: University of Chicago Press, 2011.

———. *Visible Empire: Botanical Expeditions and Visual Culture in the Hispanic Enlightenment.* Chicago: University of Chicago Press, 2011.

———. *Visual Voyages: Images of Latin American Nature from Columbus to Darwin*. New Haven, CT: Yale University Press, 2017.

Blom, Philipp. *To Have and to Hold: An Intimate History of Collectors and Collecting*. Woodstock, NY: Overlook, 2002.

Blumberg, Baruch S. "Case Study of Plant-Derived Drug Research: *Phyllanthus* and Hepatitis B Virus." In *Medicinal Plants: Their Role in Health and Biodiversity*, edited by Timothy R. Tomlinson and Akerele Olayiwola, 3–10. Philadelphia: University of Pennsylvania Press, 1998.

Blunt, Wilfrid. *The Compleat Naturalist: A Life of Linnaeus*. New York: Viking, 1971.

Blunt, Wilfrid, and William Stearn. *The Art of Botanical Illustration*. Kew, UK: Royal Botanic Gardens, Kew, 1994.

Bonpland, Aimé, Alexander von Humboldt, and Karl Sigismund Kunth. *Nova Genera et Species Plantarum*. 7 vols. Paris: Maze, 1815–25. https://www.biodiversitylibrary.org/item/270966.

Borsch, Thomas, Albert-Dieter Stevens, Eva Häffner, Anton Güntsch, Walter G. Berendsohn, Marc Appelhans, Christina Barilaro, et al. "A Complete Digitization of German Herbaria Is Possible, Sensible and Should Be Started Now." *Research Ideas and Outcomes* 6 (2020): e50675. https://doi.org/10.3897/rio.6.e50675.

Boyle, Brad, Nicole Hopkins, Zhenyuan Lu, Juan Antonio Raygoza Garay, Dmitry Mozzherin, Tony Rees, Naim Matasci, et al. "The Taxonomic Name Resolution Service: An Online Tool for Automated Standardization of Plant Names." *BMC Bioinformatics* 14, no. 1 (2013): 16. https://doi.org/10.1186/1471-2105-14-16.

Bradbury, Henry, William Grosart Johnstone, and Alexander Croall. *The Nature-Printed British Sea-weeds*. 4 vols. London: Bradbury and Evans, 1859–60. https://www.biodiversitylibrary.org/item/261169.

Brazil Flora Group. "Brazilian Flora 2020: Leveraging the Power of a Collaborative Scientific Network." *Taxon* 71, no. 1 (2022): 178–98. https://doi.org/10.1002/tax.12640.

Brockway, Lucile B. "Science and Colonial Expansion: The Role of the British Royal Botanic Gardens." *Interdisciplinary Anthropology* 6, no. 3 (1979): 449–65. https://www.jstor.org/stable/643776.

———. *Science and Colonial Expansion: The Role of the British Royal Botanic Gardens*. New York: Academic Press, 1979.

Brown, Herrick. "Catesby in Carolina." *South Carolina Wildlife*, January/February (2022): 4–11.

Brown, Jennifer, Scott E. Fulton, and Donald H. Pfister. *Glass Flowers of Harvard*. New York: Scala, 2020.

Brunfels, Otto. *Herbarum Vivae Eicones.* Strasburg: Schott, 1530. https://www.bio
    diversitylibrary.org/item/33635.

Bryant, J. A., H. Plaisier, L. M. Irvine, A. McLean, M. Jones, and M. E. Spencer
    Jones. "Life and Work of Margaret Gatty (1809–1873), with Particular Refer-
    ence to British Sea-weeds (1863)." *Archives of Natural History* 43, no. 1
    (2016): 131–47. https://doi.org/10.3366/anh.2016.0352.

Bynum, Helen, and William Bynum. *Botanical Sketchbooks.* New York: Princeton
    Architectural Press, 2017.

Cage, John. *A Mycological Foray.* Los Angeles: Atelier, 2020.

Calman, Gerta. *Ehret: Flower Painter Extraordinary.* Oxford: Phaidon, 1977.

Cámara-Leret, Rodrigo, Miguel A. Fortuna, and Jordi Bascompte. "Indigenous
    Knowledge Networks in the Face of Global Change." *Proceedings of the
    National Academy of Sciences* 116, no. 20 (2019): 9913–18. https://doi.org
    /10.1073/pnas.1821843116.

Cámara-Leret, Rodrigo, David G. Frodin, Frits Adema, Christiane Anderson,
    Marc S. Appelhans, George Argent, Susana Arias Guerrero, et al. "New
    Guinea Has the World's Richest Island Flora." *Nature* 584 (2020): 579–83.
    https://doi.org/10.1038/s41586-020-2549-5.

Campos, Paola, Clara Groot Crego, Karine Boyer, Myriam Gaudeul, C. Baider,
    Olivier Pruvost, Lionel Gagnevin, Nathalie Becker, and Adrien Rieux. "De-
    ciphering the Invasive History of a Bacterial Crop Pathogen in the Southern
    Indian Ocean Islands: Insights from Historical Herbarium Specimens."
    Conference item. *Island Biology Book of Abstracts Posters,* 2019. https://
    agritrop.cirad.fr/594130/.

Camus, Jules. "Historique des premiers herbiers." *Malpighia* 9 (1895): 286–314.

Canadensys. Accessed January 29, 2022. http://www.canadensys.net/.

Carine, Mark, ed. *The Collectors: Creating Hans Sloane's Extraordinary Herbar-
    ium.* London: Natural History Museum, London, 2020.

Carroll, Khadija von Zinnenburg, ed. *Botanical Drift: Protagonists of the Invasive
    Herbarium.* Berlin: Sternberg, 2017.

Casas, Alejandro et al. "Ethnobotany for Sustainable Ecosystem Management: A
    Regional Perspective in the Tehuacán Valley." In *Ethnobotany of Mexico: In-
    teractions of People and Plants in Mesoamerica,* edited by Rafael Lira, Ale-
    jandro Casas, and José Blancas, 179–206. New York: Springer, 2016. https://
    doi.org/10.1007/978-1-4614-6669-7_8.

Catesby, Mark. *The Natural History of Carolina, Florida and the Bahama Islands.*
    2 vols. London, 1731–43. https://www.biodiversitylibrary.org/item/41086.

Cave, Roderick. *Impressions of Nature: A History of Nature Printing.* London:
    British Library, 2010.

Čermáková, Lucie, and Jana Černá. "Naked in the Old and the New World: Differences and Analogies in Descriptions of European and American *herbae nudae* in the Sixteenth Century." *Journal of the History of Biology* 51, no. 1 (2018): 69–106. https://doi.org/10.1007/s10739-017-9468-9.

Chandra, Vikram. *Geek Sublime: The Beauty of Code, the Code of Beauty.* Minneapolis: Graywolf, 2014.

Charles, Prince of Wales. *Highgrove Florilegium.* London: Addison, 2008.

Charles, Prince of Wales, John Akeroyd, and Christopher Mills. *The Transylvania Florilegium.* 2 vols. London: Addison, 2017–19.

Chase, M. W., M. J. M. Christenhusz, M. F. Fay, J. W. Byng, W. S. Judd, D. E. Soltis, D. J. Mabberley, A. N. Sennikov, P. S. Soltis, and P. F. Stevens. "An Update of the Angiosperm Phylogeny Group Classification for the Orders and Families of Flowering Plants: APG IV." *Botanical Journal of the Linnean Society* 181 no. 1 (2016): 1–20. https://doi.org/10.1111/boj.12385.

Chase, Mark W., and James L. Reveal. "A Phylogenetic Classification of the Land Plants to Accompany APG III." *Botanical Journal of the Linnean Society* 161, no. 2 (2009): 122–27. https://doi.org/10.1111/j.1095-8339.2009.01002.x.

Childs, Arney Robinson, ed. *The Private Journal of Henry William Ravenel, 1859–1887.* Columbia: University of South Carolina Press, 1947.

Ciancio, Luca. "The Many Gardens—Real, Symbolic, Visual—of Pietro Andrea Mattioli." In *From Art to Science: Experiencing Nature in the European Garden, 1500–1700,* edited by Juliette Ferdinand, 34–45. Treviso: ZeL, 2016.

Cittadino, Eugene. *Nature as the Laboratory: Darwinian Plant Ecology in the German Empire, 1880–1900.* Cambridge: Cambridge University Press, 1990.

Clark, Anne Biller. *My Dear Mrs. Ames: A Study of Suffragist Cartoonist Blanche Ames.* New York: Lang, 2001.

Clusius, Carolus. *Exoticorum Libri Decem.* Leiden: Plantin, 1605. https://www.biodiversitylibrary.org/item/30637.

———. *Rariorum Plantarum Historia.* Antwerp: Plantin, 1601. https://www.biodiversitylibrary.org/item/14549.

Colden, Jane. *Botanic Manuscript of Jane Colden, 1724–1766.* Edited by Harold William Rickett and Elizabeth C. Hall. New York: Garden Club of Orange and Dutchess Counties, 1963.

Colella, Jocelyn P., Ryan B. Stephens, Mariel L. Campbell, Brooks A. Kohli, Danielle J. Parsons, and Bryan S. Mclean. "The Open-Specimen Movement." *BioScience* 71, no. 4 (2021): 405–14. https://doi.org/10.1093/biosci/biaa146.

Colli-Silva, Matheus, Marcelo Reginato, Andressa Cabral, Rafaela Campostrini Forzza, José Rubens Pirani, and Thais N. da C. Vasconcelos. "Evaluating Shortfalls and Spatial Accuracy of Biodiversity Documentation in the Atlantic

Forest, the Most Diverse and Threatened Brazilian Phytogeographic Domain." *Taxon* 69, no. 3 (2020): 567–77. https://doi.org/10.1002/tax.12239.

Convention on Biological Diversity. Washington, DC: Government Printing Office, 1993.

Convention on International Trade in Endangered Species of Wild Fauna and Flora | CITES, 1973. https://cites.org/eng/disc/text.php.

Cook, Alexandra. *Jean-Jacques Rousseau and Botany: The Salutary Science.* Oxford: Voltaire Foundation, 2012.

Cooper, Alix. "Placing Plants on Paper: Lists, Herbaria, and Tables as Experiments with Territorial Inventory at the Mid-Seventeenth-Century Gotha Court." *History of Science* 56, no. 3 (2018): 257–77. https://doi.org/10.1177/0073275318776515.

Corner, Edred J. H. *Botanical Monkeys.* Edinburgh: Pentland, 1992.

Cornish, Caroline, and Mark Nesbitt. "Vegetable Sheep (*Raoulia*)." In *Botanical Drift: Protagonists of the Invasive Herbarium,* edited by Khadija von Zinnenburg Carroll, 19–28. Berlin: Sternberg, 2017.

Cottesloe, Gloria, and Doris Hunt. *The Duchess of Beaufort's Flowers.* Exeter, UK: Webb and Bower, 1983.

Cowan, Richard S. "The Herbarium as a Data-bank." *Arnoldia* 33, no. 1 (1973): 3–12.

Cowell, Carly R., Pippin M. L. Anderson, and Wendy A. Annecke. "Historic Herbarium Specimens as Biocultural Assets: An Examination of Herbarium Specimens and Their in Situ Plant Communities of the Agulhas National Park, South Africa." *People and Nature* 2, no. 2 (2020): 483–94. https://doi.org/10.1002/pan3.10087.

Crane, Peter. *Ginkgo: The Tree That Time Forgot.* New Haven, CT: Yale University Press, 2013.

Crawford, Daniel J. "Plant Macromolecular Systematics in the Past 50 Years: One View." *Taxon* 49, no. 3 (2000): 479–501. https://doi.org/10.2307/1224345.

Crawford, Matthew James. *The Andean Wonder Drug: Cinchona Bark and Imperial Science in the Spanish Atlantic, 1630–1800.* Pittsburgh: University of Pittsburgh Press, 2016.

Cribb, Phillip, and Michael Tibbs. *A Very Victorian Passion: The Orchid Paintings of John Day, 1863–1888.* London: Thames and Hudson, 2004.

Cronquist, Arthur et al. *Intermountain Flora: Vascular Plants of the Intermountain West, U.S.A.* 7 vols. New York: New York Botanical Garden Press, 1972–2017.

Crowe, Victoria, and David Ingram. *Plant Memory.* Edinburgh: Royal Scottish Academy, 2007.

Culley, Theresa M. "Why Vouchers Matter in Botanical Research." *Applications in Plant Sciences* 1, no. 11 (2013): 1300076. https://doi.org/10.3732/apps.1300076.

Curtis, Simon. "The Philosopher's Flowers: John Stuart Mill as Botanist." *Encounter* 80, no. 2 (1988): 26–33.

Damasio, Antonio. *The Feeling of What Happens: Body and Emotion in the Making of Consciousness.* San Diego: Mariner, 2000.

Dandy, James Edgar. *The Sloane Herbarium.* London: British Museum, 1958. https://www.biodiversitylibrary.org/item/233790.

Daniel, Thomas F. "One Hundred and Fifty Years of Botany at the California Academy of Sciences (1853–2003)." *Proceedings of the California Academy of Sciences* 59, no. 7 (2008): 215–305.

Daniel, Thomas F., Balzac A. V. Mbola, Frank Almeda, and Peter B. Phillipson. "*Anisotes* (Acanthaceae) in Madagascar." *Proceedings of the California Academy of Sciences* 58, no. 8 (2007): 121–31. http://researcharchive.calacademy.org/research/scipubs/pdfs/v58/proccas_v58_n08.pdf.

Darlington, William. *Flora Cestrica.* West Chester, PA: Siegfried, 1837. https://www.biodiversitylibrary.org/item/122627.

Darwin, Charles. *A Monograph of the Subclass Cirripedia.* 2 vols. London: Ray Society, 1851–54. http://www.gutenberg.org/ebooks/31558.

—— *On the Origin of Species.* London: Murray, 1859. https://www.biodiversitylibrary.org/item/122307.

Darwin Correspondence Project. Cambridge University. Accessed January 29, 2022. https://www.darwinproject.ac.uk/.

Darwin, Erasmus. *The Botanic Garden.* London: Johnson, 1791.

Das, S., and M. Lowe. "Nature Read in Black and White: Decolonial Approaches to Interpreting Natural History Collections." *Journal of Natural Science Collections* 6 (2018): 4–14.

Daston, Lorraine. "Type Specimens and Scientific Memory." *Critical Inquiry* 31, no. 1 (2004): 153–82. https://doi.org/10.1086/427306.

Daston, Lorraine, and Peter Galison. *Objectivity.* New York: Zone, 2007.

Davies, Julie. "Botanizing at Badminton: The Botanical Pursuits of Mary Somerset, First Duchess of Beaufort." In *Domesticity in the Making of Modern Science,* edited by Donald Optiz, Staffan Bergwik, and Brigitte Van Tiggelen, 19–40. London: Palgrave Macmillan, 2016.

de Candolle, Alphonse. *Lois de la nomenclature botanique.* Paris: Masson, 1867. https://www.biodiversitylibrary.org/item/235808.

de Candolle, Augustin Pyramus, and Alphonse de Candolle. *Prodromus Systematis Naturalis Regni Vegetabilis.* 17 vols. Paris: Masson, 1825–73. https://www.biodiversitylibrary.org/item/7150.

Delbourgo, James. *Collecting the World: The Life and Curiosity of Hans Sloane.* Cambridge, MA: Harvard University Press, 2017.

Demissew, Sebsebe, Henk Beentje, Martin Cheek, and Ib Friis. "Sub-Saharan Botanical Collections: Taxonomic Research and Impediments." In *Tropical Plant Collections: Legacies from the Past? Essential Tools for the Future?* edited by Ib Friis and Henrik Balslev, 97–115. Copenhagen: Scientia Danica, 2015.

Desmond, Ray. *Sir Joseph Dalton Hooker: Traveller and Plant Collector.* Richmond, UK: Royal Botanic Gardens, Kew, 1999.

Dewey, John. *Art as Experience.* New York: Minton, Balch, 1934.

Diagre-Vanderpelen, Denis. "The Rise and Fall of the Belgian Forestry Museum and Geographic Arboretum (1900–1980): A Political Origin and a Winning Opportunity for Science?" *Centaurus* 60, no. 4 (2018): 333–49. https://doi.org/10.1111/1600–0498.12196.

Dickenson, Victoria. *Drawn from Life: Science and Art in the Portrayal of the New World.* Toronto: University of Toronto Press, 1998.

Dickinson, Emily. "Dickinson/Abiah Root (Strong) Correspondence: 7 May 1845 (Letter 6)." *Dickinson Electronic Archives.* Accessed January 29, 2022. http://archive.emilydickinson.org/correspondence/aroot/l6.html.

———. *Emily Dickinson's Herbarium.* Cambridge, MA: Harvard University Press, 2006.

Dieuliis, D., N. Bajema, and N. Winstead. "Biospecimens and the Information Landscape for Biodefense." Workshop. Interagency Working Group on Scientific Collections, 2019. https://wmdcenter.ndu.edu/Portals/97/Biospecimens%20and%20the%20Information%20Landscape%2010232019.pdf.

DiNoto, Andrea, and David Winter. *The Pressed Plant: The Art of Botanical Specimens, Nature Prints, and Sun Pictures.* New York: Stewart, Tabori and Chang, 1999.

Distributed System of Scientific Collections (DiSSCo). Accessed January 29, 2022. https://www.dissco.eu/.

Dobras, Werner. "Hieronymus Harder and His Twelve Plant Collections." *Ulm und Oberschwaben, Journal of History, Art and Culture* 56 (2009): 46–82.

Dold, A. P., and M. L. Cocks. "Preliminary List of Xhosa Plant Names from Eastern Cape, South Africa." *Bothalia* 29, no. 2 (1999): 267–92. https://doi.org/10.4102/abc.v29i2.601.

Doty, Mark. *Still Life with Oysters and Lemon.* Boston: Beacon, 2001.

Driessens, Erwin, and Maria Verstappen. *Herbarium Vivum.* Accessed January 29, 2022. https://notnot.home.xs4all.nl/herbariumvivum/herbariumvivum.html.

Drinkwater, Robyn. "Festival Fun with Frankenstein's Plants." *Botanics Stories* (blog), May 8, 2019. https://stories.rbge.org.uk/archives/31219.

Dupree, A. Hunter. *Asa Gray: American Botanist, Friend of Darwin.* Cambridge, MA: Harvard University Press, 1959.

Duval, Marguerite. *The King's Garden.* Charlottesville: University of Virginia Press, 1982.

Eastwood, Ruth J., Sarah Cody, Ola T. Westengen, and Roland Bothmer. "Conservation Roles of the Millennium Seed Bank and the Svalbard Global Seed Vault." In *Crop Wild Relatives and Climate Change,* edited by Robert Redden, 173–86. New York: Wiley, 2015. https://doi.org/10.1002/9781118854396.ch10.

Editorial Committee, ed. *Flora of North America North of Mexico.* 19+ vols. New York: Oxford University Press, 1993+.

Egmond, Florike. *Eye for Detail: Images of Plants and Animals in Art and Science, 1500–1630.* Chicago: University of Chicago Press, 2017.

———. "Into the Wild: Botanical Fieldwork in the Sixteenth Century." In *Naturalists in the Field: Collecting, Recording and Preserving the Natural World from the Fifteenth to the Twenty-First Century,* edited by Arthur MacGregor, 166–211. Leiden: Brill, 2018.

———. *The World of Carolus Clusius: Natural History in the Making, 1550–1610.* London: Pickering and Chatto, 2010.

Eigsti, O. J. "Induced Polypoidy." In *Fifty Years of Botany,* edited by William Campbell Steere, 153–65. New York: McGraw-Hill, 1958.

Eisenman, Sasha, Arthur Tucker, and Lena Struwe. "Voucher Specimens Are Essential for Documenting Source Material Used in Medicinal Plant Investigations." *Journal of Medicinally Active Plants* 1, no. 1 (2012): 30–43.

Elliott, Brent, Luigi Guerrini, and David Pegler. *Flora: Federico Cesi's Botanical Manuscripts.* Vol. 1. London: Royal Collection Trust, 2015.

Encyclopedia of Life. National Museum of Natural History, Smithsonian. Accessed January 29, 2022. http://eol.org/.

Endersby, Jim. *Imperial Nature: Joseph Hooker and the Practices of Victorian Science.* Chicago: University of Chicago Press, 2008.

Ent, Antony van der, Martin D. de Jonge, Rachel Mak, Jolanta Mesjasz-Przybyłowicz, Wojciech J. Przybyłowicz, Alban D. Barnabas, and Hugh H. Harris. "X-Ray Fluorescence Elemental Mapping of Roots, Stems and Leaves of the Nickel Hyperaccumulators *Rinorea* cf. *bengalensis* and *Rinorea* cf. *javanica* (Violaceae) from Sabah (Malaysia), Borneo." *Plant and Soil* 448, no. 1 (2020): 15–36. https://doi.org/10.1007/s11104-019-04386-2.

Environmental Information Reference Center (CRIA). Accessed January 29, 2022. https://www.cria.org.br/.

Erickson, Rica. *The Drummonds of Hawthornden.* Osborne Park, AUS: Lamb Paterson, 1969.

Ettingshausen, Constantin, and Alois von Pokorny. *Physiotypia Plantarum Austriacarum.* 5 vols. Vienna: Imperial Printing Office, 1855–56. https://digital.onb.ac.at/OnbViewer/viewer.faces?doc=ABO_%2BZ185917205.

Evelyn, Douglas E. "The National Gallery at the Patent Office." In *Magnificent Voyagers: The U.S. Exploring Expedition, 1838–1842,* edited by Herman J. Viola and Carolyn Margolis, 227–41. Washington, DC: Smithsonian Institution, 1985.

Evert, Ray F., and Susan Eichhorn. *Raven Biology of Plants.* 8th ed. New York: Freeman, 2014.

Ewan, Joseph. "Plant Collectors in America: Backgrounds for Linnaeus." In *Essays in Biohistory,* edited by P. Smit and R. J. ter Laage, 19–54. Utrecht, NLD: International Association for Plant Taxonomy, 1970.

———. *Rocky Mountain Naturalists.* Boulder: University of Colorado Press, 1950.

Ewan, Joseph, and Nesta Ewan. *John Banister and His Natural History of Virginia, 1678–1692.* Urbana: University of Illinois Press, 1970.

Eyde, Richard H. "Expedition Botany: The Making of a New Profession." In *Magnificent Voyagers: The U.S. Exploring Expedition, 1838–1842,* edited by Herman J. Viola and Carolyn Margolis, 25–41. Washington, DC: Smithsonian Institution Press, 1985.

Fahnestock, Jeanne. "Forming Plants in Words and Images." *Poroi* 10, no. 2 (2014): article 11. https://doi.org/10.13008/2151-2957.1194.

Fara, Patricia. *Sex, Botany, and Empire: The Story of Carl Linnaeus and Joseph Banks.* New York: Columbia University Press, 2004.

Feeley, Kenneth J., and Miles R. Silman. "The Data Void in Modeling Current and Future Distributions of Tropical Species." *Global Change Biology* 17, no. 1 (2011): 626–30. https://doi.org/10.1111/j.1365-2486.2010.02239.x.

Figueiredo, Estrela, and Gideon F. Smith. "Joaquim José da Silva (c. 1755–1810): His Life, Natural History Collecting Activities, and Involvement in the So-Called First Scientific Expedition in the Interior of Angola." *Candollea* 76, no. 1 (2021): 125–38. https://doi.org/10.15553/c2021v761a13.

Findlen, Paula. "The Death of a Naturalist: Knowledge and Community in Late Renaissance Italy." In *Professors, Physicians and Practices in the History of Medicine,* edited by Gideon Manning and Cynthia Klestinec, 127–67. New York: Springer, 2017.

———. *Possessing Nature: Museums, Collecting, and Scientific Culture in Early Modern Italy.* Berkeley: University of California Press, 1994.

Flannery, Maura C. "Flatter Than a Pancake: Why Scanning Herbarium Sheets Shouldn't Make Them Disappear." *Spontaneous Generations: A Journal of the History and Philosophy of Science* 6, no. 1 (2012): 225–32.

———. "Naming a Genus for William Darlington: A Case Study in Botanical Eponymy." *Archives of Natural History* 46, no. 1 (2019): 75–87. https://doi.org/10.3366/anh.2019.0555.

———. "Sadie Price in the Herbarium." *The Botanical Artist* 26, no. 3 (2022): 32–33.

Fleischer, Alette. "Leaves on the Loose: The Changing Nature of Archiving Plants and Botanical Knowledge." *Journal of Early Modern Studies* 6, no. 1 (2017): 117–35. https://doi.org/10.5840/jems2017616.

Flora of China Project. *Flora of China.* 44 vols. Cambridge, MA: Flora of China Project, 1983–2013.

Fraser, M., and L. Fraser. *The Smallest Kingdom: Plants and Plant Collectors at the Cape of Good Hope.* Richmond, UK: Kew Publishing, 2011.

Freer, Stephen. *Linnaeus' "Philosophia Botanica."* Oxford: Oxford University Press. 2005.

"From a Garden in Brno: The Presentation to the University of Pennsylvania of an Herbarium Specimen Prepared by Gregor Mendel." *Journal of Heredity* 27, no. 3 (1936): 91–92. https://doi.org/10.1093/oxfordjournals.jhered.a104190.

Fuchs, Leonhart. *DeHistoria Stirpium.* Basel: Isingrin, 1542. https://bibdigital.rjb.csic.es/records/item/13865-redirection.

Funk, Vicki A. "Collections-Based Science in the 21st Century." *Journal of Systematics and Evolution* 56, no. 3 (2018): 175–93. https://doi.org/10.1111/jse.12315.

———. "100 Uses for an Herbarium (Well, at Least 72)." *American Society of Plant Taxonomists Newsletter* 17, no. 2 (2003): 17–19.

Garay, Leslie. "The Orchid Herbarium of Oakes Ames: An Historical Perspective." In *Orchids at Christmas,* 41–50. Cambridge, MA: Botanical Museum of Harvard University, 2007.

Gardiner, B., and M. Morris, eds. *The Linnaean Collections.* Linnean Special Issue 7. London: Linnean Society, 2007.

Gatti, Gianna Maria. *The Technological Herbarium.* Translated by Alan N. Shapiro. Berlin: Avinus, 2010.

George, Alex. S. *William Dampier in New Holland: Australia's First Natural Historian.* Hawthorn, AUS: Bloomings Books, 1999.

Gessner, Conrad. *Historiae Animalium.* 2 vols. Zürich: Froschouer, 1551–55. https://www.biodiversitylibrary.org/item/131297.

Ghorbani, Abdolbaset, Jan J. Wieringa, Hugo J. de Boer, Henk Porck, Adriaan Kardinaal, and Tinde van Andel. "Botanical and Floristic Composition of the Historical Herbarium of Leonhard Rauwolf Collected in the Near East (1573–1575)." *Taxon* 67, no. 3 (2018): 565–80. https://doi.org/10.12705 /673.7.

Gianquitto, Tina. " 'My Dear Dr.': American Women and Nineteenth-Century Scientific Correspondence." In *The Edinburgh Companion to Nineteenth-Century American Letters and Letter-Writing,* edited by Celeste-Marie Bernier, Judie Newman, and Matthew Pethers, 435–49. Edinburgh: Edinburgh University Press, 2016.

Global Biodiversity Information Facility (GBIF). Accessed January 29, 2022. https://www.gbif.org/.

Global Plants. JSTOR. Accessed January 29, 2022. https://plants.jstor.org/.

Global Strategy for Plant Conservation. Convention on Biological Diversity. Secretariat of the Convention on Biological Diversity, 2011. https://www.cbd.int/ gspc/.

Goff, Alice. "The Schildbach Wood Library in Eighteenth-Century Hessen-Kassel." *Representations* 128, no. 1 (2014): 30–59. https://doi.org/10.1525/ rep.2014.128.1.30.

Goodman, Jordan. *Planting the World: Joseph Banks and His Collectors: An Adventurous History of Botany.* New York: HarperCollins, 2020.

Goss, Erica M., Javier F. Tabima, David E. L. Cooke, Silvia Restrepo, William E. Fry, Gregory A. Forbes, Valerie J. Fieland, Martha Cardenas, and Niklaus J. Grünwald. "The Irish Potato Famine Pathogen *Phytophthora infestans* Originated in Central Mexico Rather Than the Andes." *Proceedings of the National Academy of Sciences* 111, no. 24 (2014): 8791–96. https://doi. org/10.1073/pnas.1401884111.

Grace, Olwen M., Oscar A. Pérez-Escobar, Eve J. Lucas, Maria S. Vorontsova, Gwilym P. Lewis, Barnaby E. Walker, Lúcia G. Lohmann, et al. "Botanical Monography in the Anthropocene." *Trends in Plant Science* 26, no. 5 (2021): 433–41. https://doi.org/10.1016/j.tplants.2020.12.018.

Granados-Tochoy, Juan Carlos, Sandra Knapp, and Clara Inés Orozco. "*Solanum humboldtianum* (Solanaceae): An Endangered New Species from Colombia Rediscovered 200 Years After Its First Collection." *Systematic Botany* 32, no. 1 (2007): 200–207. https://www.jstor.org/stable/25064238.

Gray, Asa. "Diagnostic Characters of New Species of Phaenogamous Plants Collected in Japan by Charles Wright." *Memoirs of the American Academy of Arts and Sciences* 6 (1857): 377–452. https://www.biodiversitylibrary.org/ item/49960.

Gribbin, Mary, and John Gribbin. *Flower Hunters.* New York: Oxford University Press, 2008.

Gries, Corinna, Edward Gilbert, and Nico Franz. "Symbiota—A Virtual Platform for Creating Voucher-Based Biodiversity Information Communities." *Biodiversity Data Journal* 2 (2014): e1114. https://doi.org/10.3897/BDJ.2.e1114.

Gross, Alan G., and Joseph E. Harmon. *Science from Sight to Insight: How Scientists Illustrate Meaning.* Chicago: University of Chicago Press, 2013.

Gunn, Mary, and L. E Codd. *Botanical Exploration of Southern Africa.* Cape Town: Balkema, 1981.

Gutaker, Rafal M., Clemens L. Weiß, David Ellis, Noelle L. Anglin, Sandra Knapp, José Luis Fernández-Alonso, Salomé Prat, and Hernán A. Burbano. "The Origins and Adaptation of European Potatoes Reconstructed from Historical Genomes." *Nature Ecology & Evolution* 3, no. 7 (2019): 1093–1101. https://doi.org/10.1038/s41559-019-0921-3.

Haebich, Anna. "The Forgotten German Botanist Who Took 200,000 Australian Plants to Europe." *The Conversation,* July 23, 2020. http://theconversation.com/friday-essay-the-forgotten-german-botanist-who-took-200-000-australian-plants-to-europe-143099.

Hagen, Joel B. "Five Kingdoms, More or Less: Robert Whittaker and the Broad Classification of Organisms." *BioScience* 62, no. 1 (2012): 67–74. https://doi.org/10.1525/bio.2012.62.1.11.

Harbsmeier, Michael. "Fieldwork Avant La Lettre: Practicing Instructions in the Eighteenth Century." In *Scientists and Scholars in the Field: Studies in the History of Fieldwork and Expeditions,* edited by Kristian Nielsen, Michael Harbsmeier, and Christopher J. Ries, 29–50. Aarhus, DNK: Aarhus University Press, 2012.

Harkness, Deborah E. *The Jewel House: Elizabethan London and the Scientific Revolution.* New Haven, CT: Yale University Press, 2007.

Harman, Oren Solomon. *The Man Who Invented the Chromosome: A Life of Cyril Darlington.* Cambridge, MA: Harvard University Press, 2004.

Harris, Stephen. *Planting Paradise: Cultivating the Garden, 1501–1900.* Oxford: Bodleian Library, 2011.

———. *Roots to Seeds: 400 Years of Oxford Botany.* Oxford: Bodleian Library, 2021.

Hayward, Michael, and Martin Rickard. *Fern Albums and Related Material.* London: British Pteridological Society, 2019.

Heberling, J. Mason. "Herbaria as Big Data Sources of Plant Traits." *International Journal of Plant Sciences* 183, no. 2 (2022): 87–118. https://doi.org/10.1086/717623.

Heberling, J. Mason, and David J. Burke. "Utilizing Herbarium Specimens to Quantify Historical Mycorrhizal Communities." *Applications in Plant Sciences* 7, no. 4 (2019): 1–11. https://doi.org/10.1002/aps3.1223.

Heberling, J. Mason, and Bonnie L. Isaac. "Herbarium Specimens as Exaptations: New Uses for Old Collections." *American Journal of Botany* 104, no. 7 (2017): 963–65. https://doi.org/10.3732/ajb.1700125.

———. "iNaturalist as a Tool to Expand the Research Value of Museum Specimens." *Applications in Plant Sciences* 6, no. 11 (2018): e01193. https://doi.org/10.1002/aps3.1193.

Heberling, J. Mason, Joseph T. Miller, Daniel Noesgaard, Scott B. Weingart, and Dmitry Schigel. "Data Integration Enables Global Biodiversity Synthesis." *Proceedings of the National Academy of Sciences* 118, no 6 (2021): e2018093118. https://doi.org/10.1073/pnas.2018093118.

Heberling, Mason, L. Alan Prather, and Stephen Tonsor. "The Changing Uses of Herbarium Data in an Era of Global Change." *BioScience* 69, no. 10 (2019): 812–22. https://doi.org/10.1093/biosci/biz094.

Hedrick, Brandon P., J. Mason Heberling, Emily K. Meineke, Kathryn G. Turner, Christopher J. Grassa, Daniel S. Park, Jonathan Kennedy, et al. "Digitization and the Future of Natural History Collections." *BioScience* 70, no. 3 (2020): 243–51. https://doi.org/10.1093/biosci/biz163.

Helferich, Gerard. *Humboldt's Cosmos.* New York: Penguin, 2004.

Hernández, Francisco, Nardo Antonio Recchi, Joannes Terentius, Johann Faber, Fabio Colonna, and Federico Cesi. *Nova Plantarum, Animalium et Mineralium Mexicanorum Historia.* Rome: Deversini and Masotti Bibliopolarum, 1651. https://www.biodiversitylibrary.org/item/233334.

Hester, Jessica L. "Preserve Your Quarantine Nature Walks with a DIY Herbarium." *Atlas Obscura,* April 28, 2020. https://www.atlasobscura.com/articles/coronavirus-herbarium.

Hetherington, Alexander J., and Liam Dolan. "Roots in the Paleobotanical Collection." *Oxford Plant Systematics* 24 (2018): 4–5.

Hewson, Helen. *Australia: 300 Years of Botanical Illustration.* Collingwood, AUS: CSIRO, 1999.

"Hidden Treasures." *Leaflet* (Fall 2009): 24–27.

Hildreth, Jana, Eva Hrabeta-Robinson, Wendy Applequist, Joseph Betz, and James Miller. "Standard Operating Procedure for the Collection and Preparation of Voucher Plant Specimens for Use in the Nutraceutical Industry." *Analytical and Bioanalytical Chemistry* 389, no. 1 (2007): 13–17. https://doi.org/10.1007/s00216-007-1405-x.

Hill, Andrew, Robert Guralnick, Arfon Smith, Andrew Sallans, Rosemary Gillespie, Michael Denslow, Joyce Gross et al. "The Notes from Nature Tool for Unlocking Biodiversity Records from Museum Records through Citizen Science." *ZooKeys* 209 (2012): 219–33. https://doi.org/10.3897/zookeys.209.3472.

Hill, Laurence. *Fritillaria Icones.* Accessed January 29, 2022. http://www.fritillari aicones.com/.

Holmgren, Noel H. "History of the Intermountain Flora Project." In *Intermountain Flora: Vascular Plants of the Intermountain West,* edited by Arthur Cronquist et al., 7:60–73. New York: New York Botanical Garden Press, 2017.

Holstein, Norbert. "Had an Enquiry for a Single Cyperus Specimen." Twitter. *@dr_norb,* November 4, 2019. https://twitter.com/dr_norb/status/1191430 120281653250.

Holton, Gerald. *Thematic Origins of Scientific Thought: Kepler to Einstein.* Cambridge, MA: Harvard University Press, 1973.

Holway, Tatiana. *The Flower of Empire: An Amazonian Water Lily, the Quest to Make It Bloom, and the World It Created.* New York: Oxford University Press, 2013.

Hooker, Joseph Dalton. *Botany.* New York: D. Appleton, 1877. https://doi. org/10.5962/bhl.title.17868.

Hughes, Lorna. "Digital Collections as Research Infrastructure." *Educause Review,* June 2, 2014. http://www.educause.edu/ero/article/digital-collec tions-research-infrastructure.

Humboldt, Alexander von. *Cosmos.* 5 vols. London: Murray, 1849–59. https://www. biodiversitylibrary.org/item/73406#page/7/mode/1up.

———. *Essai sur la géographie des plantes.* Paris: Schoell, 2005. https://bibdigital. rjb.csic.es/idurl/1/14394.

Hunter, Dard. *Papermaking: The History and Technique of an Ancient Craft.* New York: Knopf, 1943.

iDigBio. University of Florida. Accessed January 29, 2022. https://www.idigbio.org/.

iNaturalist. Accessed January 29, 2022. https://www.inaturalist.org/.

International Image Interoperability Framework (IIIF). Accessed January 29, 2022. https://iiif.io/.

Irwin, Howard, Willard Payne, David Bates, and Philip Humphrey. *America's Systematics Collections: A National Plan.* Nebraska State Museum Report 26. Lincoln: University of Nebraska Press, 1973.

Ivins, William M. *Prints and Visual Communication.* Cambridge, MA: Harvard University Press, 1953.

Jarvis, Charlie. *Order out of Chaos: Linnaean Plant Names and Their Types.* London: Linnaean Society, 2007.

Jeppe, Carl. *Herbarium Vivum.* Rostock, DEU: Achilles, 1826.

Judson, Horace Freeland. *The Eighth Day of Creation: Makers of the Revolution in Biology.* New York: Simon and Schuster, 1979. http://archive.org/details/ eighthdayofcreatjudsrich.

Juel, H. O., and John W. Harshberger. "New Light on the Collection of North American Plants Made by Peter Kalm." *Proceedings of the Academy of Natural Sciences of Philadelphia* 81 (1929): 297–303.

Kaempfer, Engelbert. *Amoenitatum Exoticarum.* Lemgo, DEU: Meyer, 1712. https://www.biodiversitylibrary.org/item/247417.

———. *History of Japan.* London: Woodward and Davis, 1728.

Kastner, Joseph. *A Species of Eternity.* New York: Knopf, 1977.

Kingdon, Jonathan. "In the Eye of the Beholder." In *Field Notes on Science and Nature,* edited by Michael R. Canfield, 129–60. Cambridge, MA: Harvard University Press, 2011.

Kingsland, Sharon E. *The Evolution of American Ecology, 1890–2000.* Baltimore: Johns Hopkins University Press, 2005.

Kniphof, Johann Hieronymus. *Herbarium Vivum.* 12 vols. Halle, DEU: Trampe, 1758–64. https://www.biodiversitylibrary.org/item/229847.

Kohler, Robert E. *All Creatures: Naturalists, Collectors, and Biodiversity, 1850–1950.* Princeton, NJ: Princeton University Press, 2006.

Kohlstedt, Sally Gregory. "Museum Perceptions and Productions: American Migrations of a Maori Hei-Tiki." *Endeavour* 40, no. 1 (2016): 7–23. https://doi.org/10.1016/j.endeavour.2016.01.001.

Kohn, David, Gina Murrell, John Parker, and Mark Whitehorn. "What Henslow Taught Darwin." *Nature* 436 (2005): 643–45. https://doi.org/10.1038/436643a.

Koning, Jan de, Gerda van Uffelen, Alicja Zemanek, and Bogdan Zemanek, eds. *Drawn After Nature: The Complete Botanical Watercolours of the 16th-Century Libri Picturati.* Zeist, NLD: KNNV, 2008.

Kudo, Gaku, and Elisabeth J. Cooper. "When Spring Ephemerals Fail to Meet Pollinators: Mechanism of Phenological Mismatch and Its Impact on Plant Reproduction." *Proceedings of the Royal Society B: Biological Sciences* 286 (2019): 20190573. https://doi.org/10.1098/rspb.2019.0573.

Kull, Christian A., and Haripriya Rangan. "Science, Sentiment and Territorial Chauvinism in the Acacia Name Change Debate." In *Peopled Landscapes,* edited by S. G. Haberle and B. David, 197–220. Canberra: ANU-E Press, 2012.

Kusukawa, Sachiko. "Drawing as an Instrument of Knowledge: The Case of Conrad Gessner." In *Vision and Its Instruments: Art, Science and Technology in Early Modern Europe,* edited by Alina A. Payne, 36–48. University Park: Pennsylvania State University Press, 2015.

———. "Image, Text and '*Observatio*': The '*Codex Kentmanus.*'" *Early Science and Medicine* 14, no. 4 (2009): 445–75. https://www.jstor.com/stable/20617796.

———. *Picturing the Book of Nature: Image, Text, and Argument in Sixteenth-Century Human Anatomy and Medical Botany*. Chicago: University of Chicago Press, 2012.

Kuzmina, Maria L., Thomas W. A. Braukmann, Aron J. Fazekas, Sean W. Graham, Stephanie L. Dewaard, Anuar Rodrigues, Bruce A. Bennett, et al. "Using Herbarium-Derived DNAs to Assemble a Large-Scale DNA Barcode Library for the Vascular Plants of Canada." *Applications in Plant Sciences* 5, no. 12 (2017): apps.1700079. https://doi.org/10.3732/apps.1700079.

LaBouff, Nicole. "Embroidery and Information Management: The Needlework of Mary Queen of Scots and Bess of Hardwick Reconsidered." *Huntington Library Quarterly* 81, no. 3 (2018): 315–58. https://doi.org/10.1353/HLQ.2018.0014.

Lack, H. Walter. *Alexander von Humboldt and the Botanical Exploration of the Americas*. Munich: Prestel, 2009.

———. "The Botanical Field Notes Prepared by Humboldt and Bonpland in Tropical America." *Taxon* 53, no. 2 (2004): 501–10. https://doi.org/10.2307/4135629.

———. "The Plant Self Impressions Prepared by Humboldt and Bonpland in Tropical America." *Curtis's Botanical Magazine* 18, no. 4 (2001): 218–29. https://doi.org/10.1111/1467–8748.00319.

Lack, H. Walter, and David J. Mabberley. *The Flora Graeca Story: Sibthorp, Bauer, and Hawkins in the Levant*. Oxford: Oxford University Press, 1999.

LaFarge, Kelly. *Herbaria: A Guide for Young People*. St. Louis: Missouri Botanical Garden Press, 2021.

Laird, Mark. *A Natural History of English Gardening, 1650–1800*. New Haven, CT: Yale University Press, 2015.

Lanjouw, Joseph. *Index Herbariorum*. Utrecht, NLD: International Association for Plant Taxonomy, 1964.

Large, M. F., H. R. Nessia, E. K. Cameron, and D. J. Blanchon. "Changes in Stomatal Density over Time (1769–2015) in the New Zealand Endemic Tree *Corynocarpus laevigatus* J. R. Forst. & G. Forst. (Corynocarpaceae)." *Pacific Science* 71, no. 3 (2017): 319–28. https://doi.org/10.2984/71.3.6.

Latour, Bruno. "Drawing Things Together." In *Representation in Scientific Practice,* edited by Michael Lynch and Steve Woolgar, 19–68. Cambridge, MA: MIT Press, 1990.

Lavoie, Claude. "Should We Care about Purple Loosestrife? The History of an Invasive Plant in North America." *Biological Invasions* 12, no. 7 (2010): 1967–99. https://doi.org/10.1007/s10530–009–9600–7.

Lawrence, George H. M., ed. *Adanson: The Bicentennial of Michel Adanson's Familles des Plants*. Vol. 1. Pittsburgh: Hunt Botanical Library, 1963.

Leafsnap: An Electronic Field Guide. Accessed January 29, 2022. http://leafsnap.
      com/.

Leapman, Michael. *The Ingenious Mr. Fairchild: The Forgotten Father of the Flower
      Garden.* New York: St. Martin's, 2000.

Le Bras, Gwenaël, Marc Pignal, Marc L. Jeanson, Serge Muller, Cécile Aupic, Ben-
      oît Carré, Grégoire Flament, et al. "The French Muséum national d'histoire
      naturelle Vascular Plant Herbarium Collection Dataset." *Scientific Data* 4,
      no. 1 (2017): 1–16. https://doi.org/10.1038/sdata.2017.16.

Lendemer, James, Barbara Thiers, Anna K. Monfils, Jennifer Zaspel, Elizabeth R.
      Ellwood, Andrew Bentley, Katherine LeVan, et al. "The Extended Specimen
      Network: A Strategy to Enhance US Biodiversity Collections, Promote Re-
      search and Education." *BioScience* 70, no. 1 (2020): 23–30. https://doi.
      org/10.1093/biosci/biz140.

Leopold, Aldo. *A Sand County Almanac and Sketches Here and There.* New York:
      Oxford University Press, 1949.

Li, Fay-Wei, Kathleen M. Pryer, and Michael D. Windham. "*Gaga*, a New Fern
      Genus Segregated from *Cheilanthes* (Pteridaceae)." *Systematic Botany* 37,
      no. 4 (2012): 845–60. https://doi.org/10.1600/036364412X656626.

Linnaeus, Carl. *Classes Plantarum.* Leiden: Wishoff, 1738. http://linnean-online.
      org/120018/.

———. *Flora Lapponica.* Amsterdam: Schouten, 1737. http://linnean-online.
      org/120001/.

———. *Fundamenta Botanica.* Amsterdam: Schouten, 1736. http://linnean-online.
      org/119998/.

———. *Genera Plantarum.* Stockholm: Salvius, 1754. http://linnean-online.
      org/120008/.

———. *Hortus Cliffortianus.* Amsterdam, 1737. http://linnean-online.org/
      120153/.

———. *Musa Cliffortiana.* Leiden, 1736. http://www.e-rara.ch/zut/2897088.

———. *Öländska och Gothländska Resa.* Stockholm: Kiesewetter, 1745. http://
      linnean-online.org/119990/.

———. *Philosophia Botanica.* Stockholm: Kiesewetter, 1751. http://linnean-online.
      org/120027/.

———. *Species Plantarum.* 2 vols. Stockholm: Salvius, 1753. http://linnean-online.
      org/120033/.

———. *Systema Naturae.* Leiden: de Groot, 1735. https://www.biodiversitylibrary.
      org/item/15373.

Ludwig, Christian Gottlieb. *Ectypa Vegetabilium Usibus Medicis.* Halle, DEU:
      Trampe, 1760.

Luxemburg, Rosa. *Herbarium by Rosa Luxemburg.* Edited by Evelin Wittich. Berlin: Dietz, 2006.

Mabey, Richard. *The Cabaret of Plants: Forty Thousand Years of Plant Life and the Human Imagination.* New York: Norton, 2015.

MacGregor, Arthur. *Company Curiosities: Nature, Culture and the East India Company, 1600–1874.* London: Reaktion, 2018.

———. *Curiosity and Enlightenment: Collectors and Collections from the Sixteenth to the Nineteenth Century.* New Haven, CT: Yale University Press, 2007.

Mackenzie, Caitlin et al. "We Do Not Want to 'Cure Plant Blindness,' We Want to Grow Plant Love." *Plants People Planet* 1, no. 3 (2019): 139–41. https://nph.onlinelibrary.wiley.com/doi/10.1002/ppp3.10062.

Macmillan, Duncan. *Victoria Crowe: Beyond Likeness.* Edinburgh: National Galleries of Scotland, 2018.

Madriñán, Santiago. *Nikolaus Joseph Jacquin's American Plants: Botanical Expedition to the Caribbean (1754–1759) and the Publication of the "Selectarum Stirpium Americanarum Historia."* Leiden: Brill, 2013.

Mahapatra, Basudev. "Return of Indigenous Crops Helps Reduce Farm Distress and Restore Ecosystems." *Mongabay,* February 20, 2020. https://india.mongabay.com/2020/02/return-of-indigenous-crops-helps-reduce-farm-distress-and-restore-ecosystems/.

Mains, Edwin B. "Joseph Charles Arthur (1850–1942)." *Mycologia* 34, no. 6 (1942): 601–5. https://doi.org/www.jstor.org/stable/3754605.

Malthus, T. R. *An Essay on the Principle of Population.* 6th ed. 2 vols. London: Murray, 1826. https://www.biodiversitylibrary.org/item/104173.

Manzano, Saúl, and Adele C. M. Julier. "How FAIR Are Plant Sciences in the Twenty-First Century? The Pressing Need for Reproducibility in Plant Ecology and Evolution." *Proceedings of the Royal Society B: Biological Sciences* 288, no. 1944 (2021): 20202597. https://doi.org/10.1098/rspb.2020.2597.

Marder, Michael. *The Chernobyl Herbarium: Fragments of an Exploded Consciousness.* London: Open Humanities, 2016.

———. *The Philosopher's Plant: An Intellectual Herbarium.* New York: Columbia University Press, 2014.

Marsh, George Perkins. *Man and Nature, or, Physical Geography as Modified by Human Action.* London: Low, 1864. https://www.biodiversitylibrary.org/item/272449.

Marsico, Travis D., Erica R. Krimmel, J. Richard Carter, Emily L. Gillespie, Phillip D. Lowe, Ross McCauley, Ashley B. Morris, et al. "Small Herbaria Contribute Unique Biogeographic Records to County, Locality, and Temporal

Scales." *American Journal of Botany* 107, no. 11 (2020): 1577–87. https://doi.org/10.1002/ajb2.1563.

Martine, Christopher T., Emma S. Frawley, Jason T. Cantley, and Ingrid E. Jordon-Thaden. "*Solanum Watneyi,* a New Bush Tomato Species from the Northern Territory, Australia Named for Mark Watney of the Book and Film *The Martian.*" *PhytoKeys,* 61 (2016): 1–13. https://doi.org/10.3897/phytokeys.61.6995.

Masson, Veronica. "Brief History of the Collector's Index to the Wabash College Herbarium (WAB), Now Deposited at the New York Botanical Garden (NY)." *Brittonia* 46, no. 3 (1994): 211–24. https://doi.org/10.2307/2807235.

Mattioli, Pietro Andrea. *Senensis Medici.* Venice: Valgrisiana, 1565. https://www.biodiversitylibrary.org/bibliography/61850.

Mauz, Kathryn. *C. G. Pringle: Botanist, Traveler, and the "Flora of the Pacific Slope" (1881–1884).* New York: New York Botanical Garden Press, 2018.

McCarthy, Michael. *The Moth Snowstorm: Nature and Joy.* London: Murray, 2015.

McClain, Molly. *Beaufort: The Duke and His Duchess.* New Haven, CT: Yale University Press, 2001.

McKelvey, Susan Delano. *Botanical Explorations of the Trans-Mississippi West, 1790–1850.* Jamaica Plains, MA: Arnold Arboretum of Harvard University, 1955.

McKibben, Bill. *The End of Nature.* New York: Random House, 1989.

McVaugh, Rogers. *Botanical Results of the Sessé and Mociño Expedition (1787–1803).* Pittsburgh: Hunt Institute for Botanical Documentation, 2000.

Megan, Kathleen. "UConn Finds Two Plant Specimens Collected by Thoreau." *Hartford Courant,* January 15, 2011. https://www.courant.com/education/hc-xpm-2011-01-15-uconn-finds-thoreau-specimens-20110111-story.html.

Meineke, Emily K., Aimée T. Classen, Nathan J. Sanders, and T. Jonathan Davies. "Herbarium Specimens Reveal Increasing Herbivory over the Past Century." *Journal of Ecology* 107, no. 1 (2019): 105–17. https://doi.org/10.1111/1365-2745.13057.

Melo, Paula Maria Correa de Oliveira, Pedro Glécio Costa Lima, Joseane Carvalho Costa, and Márlia Regina Coelho-Ferreira. "Ethnobotanical Study in a Rural Settlement in Amazon: Contribution of Local Knowledge to Public Health Policies." *Research, Society and Development* 11, no. 1 (2022): e56911125258. https://doi.org/10.33448/rsd-v11i1.25258.

Merriman, Daniel. "Peter Artedi: Systematist and Ichthyologist." *Copeia* 1938 no. 1 (1938): 33–39. https://doi.org/10.2307/1435521.

Mickulas, Peter. *Britton's Botanical Empire: The New York Botanical Garden and American Botany, 1888–1929.* New York: New York Botanical Garden Press, 2007.

Miller, Theresa L. *Plant Kin: A Multispecies Ethnography in Indigenous Brazil.* Austin: University of Texas Press, 2019.

Milliken, William. "Mobilising Richard Spruce's 19th Century Amazon Legacy." *Read and Watch* (blog), 2017. Royal Botanic Gardens, Kew. https://www. kew.org/read-and-watch/mobilising-richard-spruce-legacy.

Miquel, Sergio E., and Frank Bungartz. "Snails Found among Herbarium Specimens of Galapagos Lichens and Bryophytes, with the Description of *Scolodonta Rinae* (Gastropoda: Scolodontidae), a New Species of Carnivorous Micro-Mollusk." *International Journal of Malacology,* 146, no. 1 (2017): 173–86. https://doi.org/10.1127/arch.moll/146/173–186.

Monardes, Nicolás. *De Simplicibus Medicamentis ex Occidentali India.* Antwerp: Plantin, 1574. https://www.biodiversitylibrary.org/item/36580.

———. *Joyfull Newes out of the Newfound Worlde.* Translated by John Frampton. London: Norton, 1580. https://www.biodiversitylibrary.org/item/31924.

Moore, Lynn M., and William K. Lauenroth. "Differential Effects of Temperature and Precipitation on Early- vs. Late-Flowering Species." *Ecosphere* 8, no. 5 (2017): e01819. https://doi.org/10.1002/ecs2.1819.

Moore, Thomas. *The Ferns of Great Britain and Ireland.* Edited by John Lindley. 2 vols. London: Bradbury and Evans, 1855. https://www.biodiversitylibrary. org/item/9497.

Moret, Pierre, Priscilla Muriel, Ricardo Jaramillo, and Olivier Dangles. "Humboldt's *Tableau Physique* Revisited." *Proceedings of the National Academy of Sciences* 116, no. 26 (2019): 12889–94. https://doi.org/10.1073/pnas.1904585116.

Morgenroth, Holly. "Research from Home: The Wild Flowers of William Keble Martin." Royal Albert Memorial Museum & Art Gallery, June 19, 2020. https://rammcollections.org.uk/2020/06/19/the-wild-flowers-of-william-keble-martin/.

Mori, Scott A. "From the Field." In *Tropical Plant Collecting,* edited by Scott A. Mori, Amy Berkov, Carol A. Gracie, and Edmund F. Hecklau, 131–90. Florianopolis, BRA: TECC Editora, 2011.

———. "My Career as a Tropical Botanist." In *Tropical Plant Collecting,* edited by Scott A. Mori, Amy Berkov, Carol A. Gracie, and Edmund F. Hecklau, 1–58. Florianopolis, BRA: TECC Editora, 2011.

Morton, Alan G. *History of Botanical Science.* New York: Academic Press, 1981.

Mucioki, Megan. "Creating Herbaria with Tribes in the Klamath River Basin." Society of Ethnobotany, April 2, 2019. https://ethnobiology.org/forage/blog/creating-herbaria-tribes-klamath-river-basin.

Mueggler, Erik. *The Paper Road: Archive and Experience in the Botanical Exploration of West China and Tibet.* Berkeley: University of California Press, 2011.

Müller-Wille, Staffan. "Linnaeus' Herbarium Cabinet: A Piece of Furniture and Its Function." *Endeavour* 30, no. 2 (2006): 60–64. https://doi.org/10.1016/j.en deavour.2006.03.001.

Müller-Wille, Staffan, and Sara Scharf. "Indexing Nature: Carl Linnaeus (1707–1778) and His Fact-Gathering Strategies." *Working Papers on the Nature of Evidence* 36 (2008): 1–39.

Mulvaney, John. *The Axe Had Never Sounded.* Canberra: ANU, 2007. www.jstor. org/stable/j.ctt24hb76.14.

Murphy, Kathleen Susan. "James Petiver's 'Kind Friends' and 'Curious Persons' in the Atlantic World: Commerce, Colonialism and Collecting." *Notes and Records: The Royal Society Journal of the History of Science* 74, no. 2 (2020): 259–74. https://doi.org/10.1098/rsnr.2019.0011.

Musgrave, T, C. Gardner, and W. Musgrave. *The Plant Hunters: Two Hundred Years of Adventure and Discovery around the World.* London: Ward Lock, 1998.

Nabhan, Gary, ed. *Ethnobiology for the Future: Linking Cultural and Ecological Diversity.* Tucson: University of Arizona Press, 2016.

———. *Where Our Food Comes From: Retracing Nikolay Vavilov's Quest to End Famine.* Washington, DC: Island, 2009.

Nabhan, Gary, and Sara St. Antoine. "The Loss of Floral and Faunal Story: The Extinction of Experience." In *The Biophilia Hypothesis,* edited by Stephen Kellert and Edward O. Wilson, 229–50. Washington, DC: Island, 1993.

The Nagoya Protocol on Access and Benefit-Sharing. Secretariat of the Convention on Biological Diversity, 2011. https://www.cbd.int/abs/.

Nagy, Amanda. "Reborn Herbarium Is a Boon to Biodiversity." *Oberlin College and Conservatory,* July 19, 2018. https://www.oberlin.edu/news/reborn-herbar ium-boon-biodiversity.

Nakamura, Tekisai, ed. *Kinmo zui.* 21 vols. Yamagata, Japan, 1666. https://dl.ndl. go.jp/info:ndljp/pid/2569340.

Nasim, Omar. "James Nasmyth's Lunar Photography; or, On Becoming a Lunar Being, without the Lunacy." In *Selene's Two Faces,* edited by Carmen González, 147–87. Leiden: Brill, 2018.

National Specimen Information Infrastructure (NSII). Accessed January 29, 2022. http://www.nsii.org.cn/2017/home.php.

"The Natural History Museum to Collaborate on Brand New Digital Project to Virtually Reunite Sir Hans Sloane's Collections." *Natural History Museum.* Accessed January 29, 2022. https://www.nhm.ac.uk/press-office/press-releases/the-natural-history-museum-to-collaborate-on-brand-new-digi tal-p.html.

Nesbitt, Mark. "Use of Herbarium Specimens in Ethnobotany." In *Curating Biocultural Collections: A Handbook,* edited by Jan Salick, Katie Konchar, and Mark Nesbitt, 313–28. Kew, UK: Royal Botanic Gardens, Kew, 2014.

Nesbitt, Mark, and Caroline Cornish. "Seeds of Industry and Empire: Economic Botany Collections between Nature and Culture." *Journal of Museum Ethnography* 29 (2016): 53–70. https://www.jstor.org/stable/43915938.

Nevill, Paul G., Xiao Zhong, Julian Tonti-Filippini, Margaret Byrne, Michael Hislop, Kevin Thiele, Stephen van Leeuwen, Laura M. Boykin, and Ian Small. "Large-Scale Genome Skimming from Herbarium Material for Accurate Plant Identification and Phylogenomics." *Plant Methods* 16, no. 1 (2020). https://doi.org/10.1186/s13007-019-0534-5.

Nicolson, Dan H. "A History of Botanical Nomenclature." *Annals of the Missouri Botanical Garden* 78, no. 1 (1991): 33–56. https://doi.org/10.2307/2399589.

Noltie, Henry J. *Botanical Art from India: The Royal Botanic Garden Edinburgh Collection.* Edinburgh: Royal Botanic Garden, Edinburgh, 2017.

———. *Indian Forester, Scottish Laird: The Botanical Lives of Hugh Cleghorn of Stravithie.* Edinburgh: Royal Botanic Garden, Edinburgh, 2016.

———. *John Hope (1725-1786): Alan G. Morton's Memoir of a Scottish Botanist.* Edinburgh: Royal Botanic Garden, Edinburgh, 2011.

"Notes from Nature." *Zooniverse.* Accessed January 29, 2022. https://www.zooniverse.org/organizations/md68135/notes-from-nature.

Novotny, Vojtech, and Kenneth Molem. "An Inventory of Plants for the Land of the Unexpected." *Nature* 584 (2020): 531–33. https://doi.org/10.1038/d41586-020-02225-4.

O'Brian, Patrick. *Joseph Banks: A Life.* Boston: Godine, 1993.

Ogilvie, Brian W. *The Science of Describing: Natural History in Renaissance Europe.* Chicago: University of Chicago Press, 2006.

Olariu, Dominic. "The Misfortune of Philippus de Lignamine's Herbal; or, New Research Perspectives in Herbal Illustrations from an Iconological Point of View." In *Early Modern Print Culture in Central Europe,* edited by Stefan Kiedron and Anna-Maria Rimm, 39–62. Warsaw: University of Warsaw Press, 2015.

Ommen, Kasper van, ed. *The Exotic World of Carolus Clusius (1526-1609).* Leiden: Leiden University Library, 2009.

O'Neill, Jean, and Elizabeth McLean. *Peter Collinson and the Eighteenth Natural History Exchange.* Philadelphia: American Philosophical Society, 2008.

"Online Floras." *World Flora Online.* Accessed January 29, 2022. https://about.worldfloraonline.org/floras.

Osseo-Asare, Abena Dove. *Bitter Roots: The Search for Healing Plants in Africa.* Chicago: University of Chicago Press, 2014.

Oviedo, Gonzalo Fernández. *Historia general de las Indias.* Seville: Cromberger, 1535. https://www.biodiversitylibrary.org/item/33562.

Pächt, Otto. "Early Italian Nature Studies and the Early Calendar Landscape." *Journal of the Warburg and Courtauld Institutes* 13, nos. 1/2 (1950): 13–47. https://doi.org/10.2307/750141.

Park, Daniel S., Xiao Feng, Shinobu Akiyama, Marlina Ardiyani, Neida Avendaño, Zoltan Barina, Blandine Bärtschi, et al. "The Colonial Legacy of Herbaria." *bioRxiv* (2021). https://doi.org/10.1101/2021.10.27.466174.

Parker, Bruce C., Noel Schanen, and Richard Renner. "Viable Soil Algae from the Herbarium of the Missouri Botanical Garden." *Annals of the Missouri Botanical Garden* 56, no. 2 (1969): 113–19. https://doi.org/10.2307/2394834.

Pavord, Anna. *The Naming of Names: The Search for Order in the World of Plants.* New York: Bloomsbury, 2005.

Pearce, Nicholas R. "John Stuart Mill's Botanical Collections from Greece (a Private Passion)." *Phytologia Balcanica* 12, no. 2 (2006): 149–64.

Pearson, Katelin D., Gil Nelson, Myla F. J. Aronson, Pierre Bonnet, Laura Brenskelle, Charles C. Davis, Ellen G. Denny, et al. "Machine Learning Using Digitized Herbarium Specimens to Advance Phenological Research." *BioScience* 70, no. 7 (2020): 610–20. https://doi.org/10.1093/biosci/biaa044.

Phelps, Almira Hart Lincoln. *Familiar Lectures on Botany.* Hartford, CT: Huntington, 1831. https://www.biodiversitylibrary.org/item/199200.

Pickering, Victoria R. M. *Putting Nature into a Box: Hans Sloane's 'Vegetable Substances' Collection.* London: University of London, 2016.

Piso, Willem, and Georg Marcgraf. *Historia Naturalis Brasiliae.* Leiden: Hackium, 1648. https://www.biodiversitylibrary.org/item/103102.

Plimpton, Pauline Ames, ed. *Oakes Ames: Jottings of a Harvard Botanist.* Cambridge, MA: Botanical Museum of Harvard University, 1979.

Polhill, Diana, and Roger Polhill. *East African Plant Collectors.* Richmond, UK: Royal Botanic Gardens, Kew, 2015.

Potter, Jennifer. *Strange Blooms: The Curious Lives and Adventures of the John Tradescants.* London: Atlantic Books, 2006.

Prather, L. Alan, Orlando Alvarez-Fuentes, Mark H. Mayfield, and Carolyn J. Ferguson. "The Decline of Plant Collecting in the United States: A Threat to the Infrastructure of Biodiversity Studies." *Systematic Botany* 29, no. 1 (2004): 15–28. https://doi.org/doi/10.1600/036364404772974185.

——— "Implications of the Decline in Plant Collecting for Systematic and Floristic Research." *Systematic Botany* 29, no. 1 (2004): 216–20. https://www.jstor.org/stable/25063950.

Prince, Sue Ann, ed. *Stuffing Birds, Pressing Plants, Shaping Knowledge: Natural History in North America, 1730–1860.* Philadelphia: American Philosophical Society, 2003.

Próćków, Jarosław, Anna Faltyn-Parzymska, Paweł Jarzembowski, Małgorzata Próćków, and Anna Jakubska-Busse. "How Many Type Specimens Can Be Stored in Old Lesser-Known Herbaria with Turbulent Histories?—A *Juncus* Case Study Reveals Their Importance in Taxonomy and Biodiversity Research." *PhytoKeys* 153 (2020): 85–110. https://doi.org/10.3897/phytokeys.153.50735.

Quinn, Stephen C. *Windows on Nature: The Great Habitat Dioramas of the American Museum of Natural History.* New York: Abrams, 2006.

Ravenel, Henry. *Fungi Caroliniani Exsiccati.* 5 vols. Charleston: Russell and Jones, 1852–60. https://digital.tcl.sc.edu/digital/collection/rav/id/9206.

Ray, John. *Historia Plantarum Generalis.* 3 vols. London: Smith and Walford, 1686–1704. https://bibdigital.rjb.csic.es/records/item/12190-redirection.

Read, Herbert. *Art and the Evolution of Man.* London: Freedom, 1951.

Riley, Margaret. "The Club at the Temple Coffee House Revisited." *Archives of Natural History* 33, no. 1 (2006): 90–100. https://doi.org/10.3366/anh.2006.33.1.90.

Robbins, Paula Ivaska. *Jane Colden: America's First Woman Botanist.* Fleischmanns, NY: Purple Mountain, 2009.

Robin, Libby, and Jane Carruthers. "National Identity and International Science: The Case of Acacia." *Historical Records of Australian Science* 23, no. 1 (2012): 34–54. https://doi.org/10.1071/HR12002.

Root-Bernstein, Robert. *Discovering.* Cambridge, MA: Harvard University Press, 1989.

Rose, Edwin D. "Natural History Collections and the Book: Hans Sloane's *A Voyage to Jamaica* (1707–1725) and His Jamaican Plants." *Journal of the History of Collections* 30, no. 1 (2018): 15–33. https://doi.org/10.1093/jhc/fhx011.

Rosser, Celia E., and Alexander S. George. *The Banksias.* 2 vols. Clayton: Monash University, 1981–88.

"Ross Potato Herbarium." *BioLib,* 1996. http://www.biolib.de/ross/potato/herbarium.html.

Roullier, C., L. Benoit, D. B. McKey, and V. Lebot. "Historical Collections Reveal Patterns of Diffusion of Sweet Potato in Oceania Obscured by Modern Plant Movements and Recombination." *Proceedings of the National Academy of Sciences* 110, no. 6 (2013): 2205–10. https://doi.org/10.1073/pnas.1211049110.

Russell, Rusty. "A Brave New Digital World." *Plant Press* 15, no. 2 (2012): 3.

Rutgers, Jorieke. "Linnaeus in the Netherlands." *Tijdschrift Voor Skandinavistiek* 29 (2008): 103–16.

Sanjad, Nelson, Ermelinda Pataca, and Rafael dos Santos. "Knowledge and Circulation of Plants: Unveiling the Participation of Amazonian Indigenous Peoples in the Construction of Eighteenth and Nineteenth Century Botany." *HoST—Journal of History of Science and Technology* 15, no. 1 (2021): 11–38. https://doi.org/10.2478/host-2021-0002.

Sargent, Matthew. "Recentering Centers of Calculation: Reconfiguring Knowledge Networks within Global Empires of Trade." In *Empires of Knowledge: Scientific Networks in the Early Modern World*, edited by Paula Findlen, 297–316. New York: Routledge, 2018.

Sassoon, Joanna. "Photographic Meaning in the Age of Digital Reproduction." *Archives and Social Studies* 1, no. 1 (2007): 299–319.

Saunders, Gill. *Picturing Plants: An Analytical History of Botanical Illustration.* Berkeley: University of California Press, 1995.

Schuyler, Alfred E., and Ann Newbold. "Vascular Plants in Lord Petre's Herbarium Collected by John Bartram." *Bartonia* 53 (1987): 41–43. https://www.jstor.org/stable/41609935.

Scurr, Ruth. *Napoleon: A Life in Gardens and Shadows.* London: Chatto and Windus, 2021.

Secord, Anne. "Corresponding Interests: Artisans and Gentlemen in Nineteenth-Century Natural History." *British Journal for the History of Science* 27, no. 4 (1994): 383–408.

SEINet Portal Network. Arizona State University. Accessed January 29, 2022. https://swbiodiversity.org/seinet/.

SERNEC: Southeast Herbaria. Appalachian State University. Accessed January 29, 2022. http://sernec.org/.

Sheldrake, Merlin. "The 'Enigma' of Richard Schultes, Amazonian Hallucinogenic Plants, and the Limits of Ethnobotany." *Social Studies of Science* 50, no. 3 (2020): 345–76. https://doi.org/10.1177/0306312720920362.

Shepherd, Lara D., Marlies Thiedemann, and Carlos Lehnebach. "Genetic Identification of Historic *Sophora* (Fabaceae) Specimens Suggests Toromiro (*S. toromiro*) from Rapa Nui/Easter Island May Have Been in Cultivation in Europe in the 1700s." *New Zealand Journal of Botany* 58, no. 3 (2020): 255–67. https://doi.org/10.1080/0028825X.2020.1725069.

Sherwood, Shirley. *Contemporary Botanical Artists.* New York: Cross River, 1996.

Sherwood, Shirley, Stephen A. Harris, and B. E. Juniper. *A New Flowering: 1000 Years of Botanical Art.* Oxford: Ashmolean, 2005.

Shetler, Stanwyn. "The Herbarium: Past, Present, and Future." *Proceedings of the Biological Society of Washington* 82 (1969): 687–758. https://www.biodiver sitylibrary.org/part/43553.

———. *An Introduction to the Botanical Type Specimen Register.* Washington, DC: Smithsonian Institution Press, 1973. https://www.biodiversitylibrary.org/ bibliography/123274.

Silvertown, Jonathan. *Demons in Eden: The Paradox of Plant Diversity.* Chicago: University of Chicago, 2005.

Simpson, Niki, and Peter G. Barnes. "Photography and Contemporary Botanical Illustration." *Curtis's Botanical Magazine* 25, no. 3 (2008): 258–80. https:// doi.org/10.1111/j.1467-8748.2008.00628.x.

Sivasundaram, Sujit. "The Oils of Empire." In *Worlds of Natural History,* edited by H. A. Curry, N. Jardine, J. A. Secord, and E. C. Spary, 379–98. Cambridge: Cambridge University Press, 2018.

Sloane, Hans. *Natural History of Jamaica.* 2 vols. London, 1707–25. https://www. biodiversitylibrary.org/item/11242.

Smith, Bernard. *European Vision and the South Pacific.* 2nd ed. New Haven, CT: Yale University Press, 1985.

Smith, Gideon F. "The African Plants Initiative: A Big Step for Continental Taxonomy." *Taxon* 53, no. 4 (2004): 1023–25. https://doi.org/10.2307/4135568.

Smith, Pamela H. *The Body of the Artisan: Art and Experience in the Scientific Revolution.* Chicago: University of Chicago Press, 2003.

Spamer, Earle, and Richard McCourt. "The Lewis and Clark Herbarium of the Academy of Natural Sciences, Part 1." *Notulae Naturae* 475 (2002): 1–46.

Spellman, Katie V., and Christa P. H. Mulder. "Validating Herbarium-Based Phenology Models Using Citizen-Science Data." *BioScience* 66, no. 10 (2016): 897–906. https://doi.org/10.1093/biosci/biw116.

Stafleu, Frans A. "Benjamin Delessert and Antoine Lasegue." *Taxon* 19, no. 6 (1970): 920–38. https://doi.org/10.2307/1218312.

Star, Susan Leigh, and James R. Griesemer. "Institutional Ecology, 'Translations' and Boundary Objects: Amateurs and Professionals in Berkeley's Museum of Vertebrate Zoology, 1907–39." *Social Studies of Science* 19, no. 3 (1989): 387–420. https://www.jstor.org/stable/285080.

Stearn, W. T. Introduction to *Species Plantarum: A Facsimile of the First Edition, 1753,* 1:1–199. London: Ray Society, 1957.

———. "Mrs. Agnes Arber, Botanist and Philosopher, 1879–1960." *Taxon* 9, no. 9 (1960): 261–63. https://www.jstor.org/stable/1217828.

Stearns, Raymond P. "James Petiver: Promoter of Natural Science, c. 1663–1718." *Proceedings of the American Antiquarian Society* 62 (1952): 243–365.

——. *Science in the British Colonies of America.* Urbana: University of Illinois Press, 1970.

Steele, Arthur Robert. *Flowers for the King: The Expedition of Ruiz and Pavon and the Flora of Peru.* Durham, NC: Duke University Press, 1964.

Stefanaki, Anastasia, Henk Porck, Ilaria Maria Grimaldi, Nikolaus Thurn, Valentina Pugliano, Adriaan Kardinaal, Jochem Salemink, et al. "Breaking the Silence of the 500-Year-Old Smiling Garden of Everlasting Flowers: The En Tibi Book Herbarium." *PLOS ONE* 14, no. 6 (2019): e0217779. https://doi.org/10.1371/journal.pone.0217779.

Stepp, J. R., and M. B. Thomas. "Managing Ethnopharmacological Data: Herbaria, Relational Databases, Literature." *Medical and Health Sciences* 13 (2010): 116–23.

Stevens, Peter F. *The Development of Biological Systematics: Antoine-Laurent de Jussieu, Nature, and the Natural System.* New York: Columbia University Press, 1994.

Stevenson, Jan Wassmer, and Dennis Wm. Stevenson. "The Nuts and Bolts of Doing the *Flora of the Bahama Archipelago*: How Don Correll Worked." *Botanical Review* 80, no. 3 (2014): 135–47. https://doi.org/10.1007/s12229-014-9144-0.

Stoffelen, Piet et al. "Central African Plants," 2009. http://projects.biodiversity.be/brcap/pages/about.

Strasser, Bruno J. "The Experimenter's Museum: GenBank, Natural History, and the Moral Economics of Biomedicine." *Isis* 102, no. 1 (2011): 60–96. https://doi.org/10.1086/658657.

"A Strategic Plan for Establishing a Network Integrated Biocollections Alliance," 2010. https://digbiocol.files.wordpress.com/2010/08/niba_brochure.pdf.

Strocchia, Sharon T. *Forgotten Healers: Women and the Pursuit of Health in Late Renaissance Italy.* Cambridge, MA: Harvard University Press, 2019.

Stuckey, Ronald L. "The First Public Auction of an American Herbarium, Including an Account of the Fate of the Baldwin, Collins, and Rafinesque Herbaria." *Taxon* 20, no. 4 (1971): 443–59. https://doi.org/10.2307/1218245.

Stungo, Ruth. "The Royal Society Specimens from the Chelsea Physic Garden, 1722–1799." *Notes and Records of the Royal Society of London* 47, no. 2 (1993): 213–24. https://doi.org/10.1098/rsnr.1993.0028.

Teltscher, Kate. *Palace of Palms: Tropical Dreams and the Making of Kew.* London: Picador, 2020.

Thiers, Barbara. *Herbarium: The Quest to Preserve and Classify the World's Plants.* Portland, OR: Timber, 2020.

——. *Index Herbarorum.* New York Botanical Garden. Accessed January 29, 2022. http://sweetgum.nybg.org/science/ih/.

———. *The World's Herbaria, 2020: A Summary Report Based on Data from Index Herbariorum.* New York: New York Botanical Garden, 2021.

Thiers, Barbara et al. "Extending U.S. Biodiversity Collections to Promote Research and Education." *Biodiversity Collections Network,* 2019. https://www.idigbio.org/sites/default/files/sites/default/files/BCoN/Extending-Biodiversity-Collections-Full-Report%282%29.pdf.

Thijsse, Gerard. "A Contribution to the History of the Herbaria of George Clifford III (1685–1760)." *Archives of Natural History* 45, no. 1 (2018): 134–48. https://doi.org/10.3366/anh.2018.0489.

Thinard, Florence. *Explorers' Botanical Notebook.* Buffalo, NY: Firefly, 2016.

Thomas, Ann. *Beauty of Another Order: Photography in Science.* New Haven, CT: Yale University Press, 1997.

Timby, Sara. "The Dudley Herbarium, Including a Case Study of Terman's Restructuring of the Biology Department." *Sandstone and Tile* 22, no. 4 (1998): 3–15.

Tognoni, Federico. "Nature Described: Fabio Colonna and Natural History Illustration." *Nuncius* 20, no. 2 (2005): 347–70. https://doi.org/10.1163/182539105X00024.

Tomasi, Lucia Tongiorgi. *An Oak Spring Flora: Flower Illustration from the Fifteenth Century to the Present Time.* Upperville, VA: Oak Spring Garden Library, 1997.

Tomasi, Lucia Tongiorgi, and Tony Willis. *An Oak Spring Herbaria: Herbs and Herbals from the Fourteenth to the Nineteenth Centuries.* Upperville, VA: Oak Spring Garden Library, 2009.

Toresella, Sergio, and Marisa Battini. "Gli erbari a impressione e l'origine del disegno scientifico." *Le scienze* 239, no. 7 (1988): 64–78.

Torrey, John. *On the Darlingtonia Californica, a New Pitcher-Plant from Northern California.* Washington, DC: Smithsonian Institution, 1853. https://www.biodiversitylibrary.org/item/15291.

Tournefort, Joseph Pitton de. *Élémens de botanique.* 3 vols. Paris: L'Imprimerie Royale, 1694. https://bibdigital.rjb.csic.es/records/item/13697-redirection.

Tribe, H. "The Dillon-Weston Glass Models of Microfungi." *Mycologist* 31 (1998): 169–73.

Turland, Nicholas, John Wiersema, Fred Barrie, Werner Greuter, David Hawksworth, Patrick Herendeen, Sandra Knapp, et al., eds. *International Code of Nomenclature for Algae, Fungi, and Plants.* Glashütten, DEU: Koeltz, 2018. https://doi.org/10.12705/Code.2018.

Ubrizsy Savoia, Andrea. "The Influence of New World Species on the Botany of the 16th Century." *Asclepio* 48, no. 2 (1996): 163–72. https://doi.org/10.3989/asclepio.1996.v48.i2.403.

Václavík, Tomáš, Michael Beckmann, Anna F. Cord, and Anja M. Bindewald. "Effects of UV-B Radiation on Leaf Hair Traits of Invasive Plants—Combining Historical Herbarium Records with Novel Remote Sensing Data." *PLOS ONE* 12, no. 4 (2017): e0175671. https://doi.org/10.1371/journal. pone.0175671.

van Rheede tot Drakestein, Hendrik Adriaan. *Hortus Indicus Malabaricus.* 12 vols. Amsterdam: van Someren and van Dyck, 1678–1703. https://bibdigital.rjb. csic.es/records/item/14027-redirection.

Vines, Sydney Howard. "Robert Morison, 1620–1683, and John Ray, 1627–1705." In *Makers of British Botany,* edited by F. W. Oliver, 8–43. Cambridge: Cambridge University Press, 1913.

Voeks, Robert A. *The Ethnobotany of Eden: Rethinking the Jungle Medicine Narrative.* Chicago: University of Chicago Press, 2018.

Wallich, Nathaniel. *Plantae Asiaticae Rariores.* 3 vols. London: Treuttel and Würtz, 1830–32. https://doi.org/1r8.

Watt, Alistair. *Robert Fortune: A Plant Hunter in the Orient.* Richmond, UK: Kew Publishing, 2017.

Webb, Joan. *The Botanical Endeavor: Journey towards a Flora of Australia.* Chipping Norton, AUS: Surrey Beatty, 2003.

Weiss, Amy. "Art Cronquist's Hat." *Hand Lens* (blog). NYBG Steere Herbarium, May 9, 2019. http://sweetgum.nybg.org/science/the-hand-lens/explore/ narratives-details/?irn=6993.

White, Paul. "The Purchase of Knowledge: James Edward Smith and the Linnaean Collections." *Endeavor* 23, no. 3 (1999): 126–29. https://doi.org/10.1016/ S0160-9327(99)01212-0.

Whitehead, D. R. "Collecting Beetles in Exotic Places: The Herbarium." *Coleopterists Bulletin* 30, no. 3 (1976): 249–50.

Whitfield, John. "Superstars of Botany: Rare Specimens." *Nature* 484 (2012): 436–38. https://doi.org/10.1038/484436a.

Williams, Glyn. *Naturalists at Sea: Scientific Travelers from Dampier to Darwin.* New Haven, CT: Yale University Press, 2013.

Williams, Roger L. *French Botany in the Enlightenment: The Ill-Fated Voyages of La Pérouse and His Rescuers.* Dordrecht, NLD: Kluwer, 2003.

Willis, Kathy J. *State of the World's Fungi, 2018.* Richmond, UK: Royal Botanic Gardens, Kew, 2018.

Wilson, Edward O. *Biophilia.* Cambridge, MA: Harvard University Press, 1984.

Wilson, Edward O., and Frances Peter, eds. *Biodiversity.* Washington, DC: National Academy Press, 1988.

Winston, Judith E. *Describing Species: Practical Taxonomic Procedure for Biologists.* New York: Columbia University Press, 1999.

Winterbottom, Anna. "Medicine and Botany in the Making of Madras, 1680–1720." In *The East India Company and the Natural World,* edited by V. Damodaran, Anna Winterbottom, and Alan Lister, 35–57. London: Palgrave Macmillan, 2015.

Wolkis, Dustin, Kelli Jones, Tim Flynn, Mike DeMotta, and Nina Rønsted. "Germination of Seeds from Herbarium Specimens as a Last Conservation Resort for Resurrecting Extinct or Critically Endangered Hawaiian Plants." *Conservation Science and Practice* 4, no. 1 (2022): e576. https://doi.org/10.1111/csp2.576.

World Flora Online. Accessed January 29, 2022. http://www.worldfloraonline.org/.

Wulf, Andrea. *The Brother Gardeners: Botany, Empire and the Birth of an Obsession.* New York: Knopf, 2009.

———. *The Invention of Nature: Alexander von Humboldt's New World.* New York: Knopf, 2015.

Yang, Andrew. "That Drunken Conversation between Two Cultures: Art, Science and the Possibility of Meaningful Uncertainty." *Leonardo* 48, no. 3 (2015): 318–21. https://doi.org/10.1162/LEON_a_00705.

Yost, Jenn M., Katelin D. Pearson, Jason Alexander, Edward Gilbert, Layla Aerne Hains, Teri Barry, Robin Bencie, et al. "The California Phenology Collections Network: Using Digital Images to Investigate Phenological Change in a Biodiversity Hotspot." *Madroño* 66, no. 4 (2020): 130–41. https://doi.org/10.3120/0024-9637-66.4.130.

Young, Donna. "Brendel Plant Model Survey." *NatSCA* (blog), August 1, 2019. https://natsca.blog/2019/08/01/brendel-plant-model-survey/.

Zangerl, Arthur, and May Berenbaum. "Increase in Toxicity of an Invasive Weed After Reassociation with Its Coevolved Herbivore." *Proceedings of the National Academy of Sciences* 102, no. 43 (2005): 15529–32. https://doi.org/10.1073/pnas.0507805102.

Ziman, J. M. *Public Knowledge.* Cambridge: Cambridge University Press, 1968.

Zoller, Heinrich, Martin Steinmann, and Karl Schmid. *Conradi Gesneri Historia Plantarum.* Facsimile. 8 vols. Zürich: Urs Graf-Verlag, 1972–80.

Zooniverse. Accessed May 8, 2022. https://www.zooniverse.org/.

Index

Figures are indicated by "f" following the page numbers.